# Modeling and Flowsheet Simulation of Vibrated Fluidized Bed Dryers

# Modeling and Flowsheet Simulation of Vibrated Fluidized Bed Dryers

Vom Promotionsausschuss der
Technischen Universität Hamburg
zur Erlangung des akademischen Grades

**Doktor-Ingenieur (Dr.-Ing.)**

genehmigte Dissertation

von

**Sören Ernst Lehmann**
aus
Buchholz i.d.N.

2021

**Bibliografische Information der Deutschen Nationalbibliothek**
Die Deutsche Nationalbibliothek verzeichnet diese Publikation in der
Deutschen Nationalbibliografie; detaillierte bibliographische Daten sind im Internet
über http://dnb.d-nb.de abrufbar.
1. Aufl. - Göttingen: Cuvillier, 2021
Zugl.: (TU) Hamburg, Univ., Diss., 2021

*Gutachter:*
1. Prof. Dr.-Ing. habil. Dr. h.c. Stefan Heinrich
2. Prof. Dr.-Ing Michael Schlüter

*Vorsitz:*
Prof. Dr. Raimund Horn

*Tag der mündlichen Prüfung:*
17.11.2021


Nonnenstieg 8, 37075 Göttingen
Telefon: 0551-54724-0
Telefax: 0551-54724-21
www.cuvillier.de


1. Auflage, 2021
Gedruckt auf umweltfreundlichem, säurefreiem Papier aus nachhaltiger Forstwirtschaft.

ISBN 978-3-7369-7539-2
eISBN 978-3-7369-6539-3

# Acknowledgment

The list of people I have to thank for their parts within the last 4,5 years. Foremost Prof. Stefan Heinrich for the opportunity to work on this interesting topic and allow me to take on my journey in research. Also, Alfred Jongsam and Fredrik Innings from TetraPak for initiating the project and giving valuable advice and providing industrial and application perspectives. It was a pleasure working with you.
Specifically, I want mention Ernst-Ulrich Hartge, for his support and guidance during the first half of my journey. He passed away far too early and is missed dearly.
Special thanks go out to Swantje Pietsch-Braune for proof-reading and to Moritz Buchholz for his immeasurable assistance and support with programming and simulation issues, to Heiko Rhode and and Frank Rimoschat for their support and advice in design, construction and operation of the experimental facilities and of course to all the colleagues at the SPE, who became friends along the way. All of you made this time interesting, fun and unforgettable.
Last but certainly not least, I thank all the students who assisted during the experiments and spent countless hours in the labs for this project: Tobias Thiele, Paramveer Gopalsingh, Diego Espinoza, Arun Nair, Oshin Sen, Brandon Boepple, Deifila To, Xuebin Feng, Hanna Reichenbach, Imen Barrack, Marius Fiedler, Raoul Sonkoue, Zhengyu Lu, Tobias Oesau, Justin Dittmer, Lukas Frank and Dominik Kwidzinski.

# *Abstract*

**Modeling and Flowsheet Simulation of Vibrated Fluidized Bed Dryers**

by Sören Ernst LEHMANN

Fluidized bed dryers are the prime choice when it comes to drying of heat sensitive products, commonly processed in the pharmaceutical and food industry. As many products in these industries are fine and cohesive, mechanical vibration of the dryer is used to enable or improve fluidization. Thus, the goal of this thesis is the development of a fluidized bed drying model that accounts for the influence of mechanical vibration of the dryer, as well as its implementation in an open-source flowsheet simulation framework. Continuously operated fluidized bed dryers under steady-state conditions are the focus of this thesis. The aim during model development and implementation is the broadest possible application range of the model. A custom-built vibrated fluidized bed dryer is designed and constructed for comprehensive investigations of fluidized bed hydrodynamics and drying kinetics. Based on experimental investigations, a semi-empirical model for hydrodynamics of fine and cohesive powders is developed. The new model is combined with established models to allow for the flowsheet simulation of fluidized bed dryers for particles of all Geldart groups. Additionally, the influence of vibration is accounted for. Comprehensive validation experiments are performed for particles of different Geldart groups, different dryer geometries and a variety of process parameters, including mechanical vibration. Comparison of model predictions with experimental data attributes high accuracy of predicted particle and gas properties. Furthermore, sensitivity analyses are conducted to identify potential weaknesses in underlying model assumptions. Hereby, the validity of underlying assumptions is confirmed and potential optimization parameters for different applications are identified. The proposed model is unprecedented in terms of range of process parameters, variety of particle properties and dryer geometries, tested and found valid for.

# Contents

# Symbols

## Latin Symbols

| | | |
|---|---|---|
| $A$ | surface or cross-sectional area | $m^2$ |
| $A_{vib}$ | amplitude of vibration | m |
| $Ar$ | Archimedes number | – |
| $Bi$ | Biot number | – |
| $Bo$ | Bodenstein number | – |
| $c$ | concentration | $kg\,m^{-3}$ |
| $c_p$ | heat capacity | $J\,kg^{-1}\,K^{-1}$ |
| $d$ | diameter | m |
| $D$ | diffusion coefficient | $m\,s^{-1}$ |
| $\Delta E_V$ | activation energy | J |
| $\Delta h_V$ | evaporation enthalpy | $J\,kg^{-1}$ |
| $f$ | frequency of vibration | Hz |
| $g$ | gravitational acceleration | $m\,s^{-2}$ |
| $h$ | specific enthalpy | $J\,kg^{-1}$ |
| $H$ | height | m |
| $\dot{H}$ | enthalpy flow | W |
| $k$ | relative bubble velocity ratio | – |
| $K$ | relative dielectric constant | – |
| $K$ | number of tanks in series | – |
| $l$ | modified free length of gas molecule | m |
| $Le$ | Lewis number | – |
| $\dot{m}$ | mass flux | $kg\,s^{-1}\,m^{-2}$ |
| $M$ | mass | kg |
| $\dot{M}$ | mass flow | $kg\,s^{-1}$ |
| $\tilde{M}$ | molar mass | $mol\,g^{-1}$ |
| $MR$ | moisture ratio | – |
| $n$ | number density distribution of residence time | – |
| $N$ | total number | – |
| $N_A$ | Avogadro constant | $6.022 \times 10^{-23}\,mol^{-1}$ |

| | | |
|---|---|---|
| $Nu$ | Nusselt number | – |
| $p$ | pressure | Pa |
| $Pe$ | Peclet number | – |
| $Pr$ | Prandtl number | – |
| $\dot{Q}$ | heat flow | W |
| $R$ | gas constant | $\mathrm{J\,kg^{-1}\,K}$ |
| $Re$ | Reynolds number | – |
| $S$ | specific surface area | – |
| $Sc$ | Schmidt number | – |
| $Sh$ | Sherwood number | – |
| $t$ | time | s |
| $T$ | temperature | K |
| $u$ | velocity | $\mathrm{m\,s^{-1}}$ |
| $\dot{V}$ | volume flow rate | $\mathrm{m^3\,s^{-1}}$ |
| $\dot{V}_B$ | visible bubble flow rate | $\mathrm{m\,s^{-1}}$ |
| $X$ | moisture content of solids | $\mathrm{kg\,kg^{-1}}$ |
| $Y$ | gas moisture content | $\mathrm{kg\,kg^{-1}}$ |
| $z$ | height above distributor plate | m |

## Greek Symbols

| | | |
|---|---|---|
| $\alpha$ | heat transfer coefficient | $\mathrm{J\,kg^{-1}\,K^{-1}}$ |
| $\beta$ | mass transfer coefficient | $\mathrm{m\,s^{-1}}$ |
| $\varepsilon$ | porosity | – |
| $\zeta$ | normalized bed height | – |
| $\eta$ | dynamic viscosity | $\mathrm{Pa\,s^{-1}}$ |
| $\vartheta$ | temperature | °C |
| $\theta$ | dimensionless residence time | – |
| $\Theta$ | hydrodynamic parameter | – |
| $\kappa$ | bubble coalescence dynamics factor | – |
| $\lambda$ | average bubble life time | s |
| $\lambda_G$ | heat conductivity of gas | W/m/K |
| $\Lambda$ | vibration intensity | – |
| $\nu$ | kinematic viscosity | $\mathrm{m^2\,s}$ |
| $\dot{\nu}$ | material specific drying curve | – |
| $\xi$ | normalized moisture content | – |
| $\rho$ | density | $\mathrm{kg\,m^{-3}}$ |
| $\tau$ | residence time | s |
| $\tilde{\tau}$ | mean residence time | s |
| $\bar{\tau}$ | hydrodynamic residence time | s |
| $\varphi$ | relative humidity | – |

| | | |
|---|---|---|
| $\Phi$ | relative humidity at particle surface | – |
| $\chi$ | dimensionless axial coordinate | – |
| $\psi$ | hydrodynamic parameter | – |
| $\Psi_{\mathrm{Wadell}}$ | sphericity after Wadell | – |
| $\omega$ | Weibull extrapolation factor | – |

## Subscripts

| | |
|---|---|
| 0 | initial or reference state |
| $\infty$ | surrounding condition |
| amb | ambient |
| app | apparent |
| ax | axial |
| B | bubble phase |
| cham | chamber |
| d | discretized particle size class |
| cr | critical |
| eq | equilibrium |
| exp | experiment |
| f | discretized particle moisture class |
| fb | fluidized bed under operation conditions |
| fix | fixed bed |
| G | gas phase |
| i | current step |
| in | inlet |
| lam | laminar |
| m | adsorbed molecules |
| max | maximum |
| mf | minimum fluidization |
| op | operational |
| or | orifice |
| out | outlet |
| P | particle phase / solid phase |
| pf | solid free |
| ref | reference |
| res | resonance |
| s | surface |
| S | suspension phase |
| sat | saturation |
| sim | simulation |
| sk | skeletal |

| | |
|---|---|
| surf | surface equivalent |
| t | terminal |
| T | tracer |
| turb | turbulent |
| v | vapor/gaseous |
| vol | volume equivalent |
| W | wall |
| wb | wet bulb |

## Abbreviations

| | |
|---|---|
| ABF | aggregate bubble fluidization |
| APF | aggregate particle fluidization |
| CDC | Characteristic Drying Curve |
| CSTR | continuous stirred tank reactor |
| DP | drying period |
| DPM | distributed parameter model |
| DVS | dynamic vapor sorption |
| DYSSOL | Dynamic Simulation of SOLids processes |
| fps | frames per second |
| FCC | fluid catalytic cracking |
| GUI | graphical user interface |
| IUPAC | International Union of Pure and Applied Chemistry |
| LPM | lumped parameter model |
| NCDC | Normalized Characteristic Drying Curve |
| NTU | number of transfer units |
| PFR | plug flow reactor |
| REA | Reaction Engineering Approach |
| RTD | residence time distribution |
| SPE | Institute of Solids Process Engineering and Particle Technology |
| TIS | tank in series |
| TUHH | Hamburg University of Technology |
| VFB | vibrated fluidized bed |
| WMP | whole milk powder |

# 1

# Introduction

## 1.1 Fluidized Bed Technology

The first industrial application of fluidized bed technology was documented in Germany in 1922. The application was the production of hydrogen gas from fuel particles, suspended in a fluid-like state by a gas flow (Winkler, 1922). This fluid-like or fluidization state is reached, when gravitational and inertia forces of particles are in equilibrium with drag forces, exercised by an upwards-flowing gas. In fluidized beds, the particles are constantly in motion, resulting in high degree of mixing. Also, the porosity of the bed is increased, compared to the fixed bed state. Both lead to significantly higher heat and mass transfer between particles and fluidization gas. The mixing additionally entails homogeneous particle properties, such as temperature, moisture content, or progress of potential chemical reactions with the gas phase (Kunii et al., 2013).

Nowadays, fluidized beds are widely applied in several industrial fields, such as chemical engineering, combustion, mineral and ore processing, or pharmaceutical and food industry. Applications vary from drying purposes over chemical reactions to particle formulation. Regardless of the application, the interactions between the fluidization gas and the particles determine the properties and potential uses of the fluidized bed (Werther et al., 2014). Those interactions are summarized under the term hydrodynamics, which covers the flow of the gas in form of suspension gas and bubbles, the size, velocity and flow pattern of gas bubbles, the mixing and flow pattern of the particles as well as the resulting bed expansion, respectively bed porosity. Different flow regimes are achieved by variation of gas flow rate (Kunii et al., 2013). The most important flow regimes of gas solid fluidized beds are shown in Figure 1.1.

The fixed bed regime (A) is the regime where the gas, passing through the bed, does not apply sufficient drag to set the particles in motion. Once the drag forces on the particles are large enough to lift the particles and suspend them, minimum fluidization is reached (B). The onset of fluidization, also called minimum fluidization velocity $u_{\mathrm{mf}}$, marks the theoretical lower end of the operational range of fluidized beds. When the gas velocity is increased, the pressure drop across the bed stays constant until particles are entrained

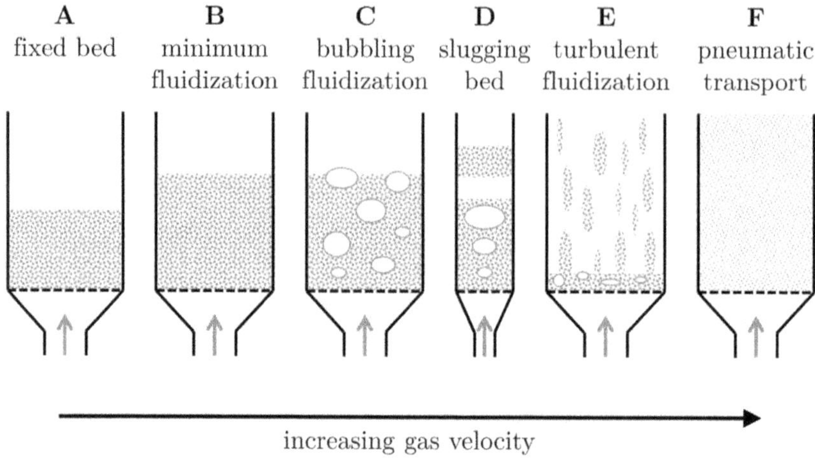


Figure 1.1: Relevant flow regimes of gas solid fluidized beds, depending on gas velocity.

from the bed. This is accompanied by expansion of the bed. Further increase in gas velocity results in the formation of bubbles. The bubbling regime (C) is accompanied by further expansion of the bed. The bubbles cause intense mixing of the particles, which results in the high heat and mass transfer rates, characteristic for fluidized beds. In case the bed is narrow, bubbles may grow as large as the cross-sectional area of the bed, resulting in slugging (D). For further increased gas velocity, the turbulent regime (E) follows. Here, the particles are moving upwards in strains or small clusters and travel downwards near the walls. The onset of particles being dragged from the bed is called the elutriation velocity $u_{\mathrm{elu}}$. Operation of fluidized beds at or above $u_{\mathrm{elu}}$ is applied in some cases, e.g. chemical looping combustion or catalytic cracking (Kunii et al., 2013), but are beyond the scope this work and are therefore not further discussed. When the gas velocity is increased even further, pneumatic transport (F) of individual particles is achieved, the particles are carried out of the system. This work focuses on bubbling fluidized beds.

The so-called *two-phase model* is used to describe the hydrodynamics of fluidized beds. It states that the gas can be distinguished into two phases: 1) the bubbles traveling through the bed, which are mainly responsible for mixing the particles; as well as 2) the suspension gas, which is in direct contact with the particles and is thus, responsible for most of the heat and mass transfer between particles and gas (Davidson and Harrison, 1966). The behavior of fluidized beds depends on particle properties. The most common characterization was introduced by Geldart (1973). Therein, the fluidization behavior of particles is classified by the mean particle size and density difference between particles and fluidization gas, resulting in four Geldart groups. Molerus (1982) refined Geldart's empirical findings by investigating the boundary regions between different Geldart groups. He defined the boundaries between the groups based on force

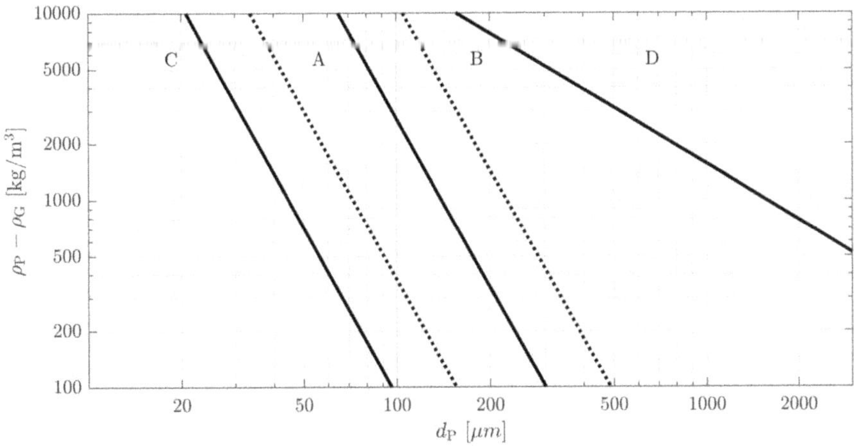


Figure 1.2: Molerus diagram, showing Geldart groups based on particle properties and respective boundaries, derived by force balances. Solid lines depict boundaries for hard particles, dashed lines indicate boundaries for softer particles, showing higher interparticle cohesiveness, according to Molerus (1982).

balances, accounting for van der Waals forces, cohesiveness due to particle hardness and the drag force exerted by the fluidization gas. The classification of powders in Geldart groups after Molerus (1982) is shown Figure 1.2.

Fluidized beds of Geldart group A expand at $u_{mf}$ before bubbling occurs with increasing gas velocity. The bubbles are categorized as fast bubbles, because they move faster through the bed than the suspension gas. Given sufficient bed height, an equilibrium between growth and splitting of bubbles is reached, leading to a maximum bubble size. The bed collapses slowly when the gas stream is suddenly turned off (Geldart, 1973).

Group B particles have larger mean diameters and densities than group A particles. Bubbling starts immediately at $u_{mf}$. The bubbles also travel faster than the suspension gas. However, the bubble size does not reach an equilibrium. Bubble growth in the bed is only limited by the apparatus walls. Thus, slugging of the bed can occur with group B particles. When the gas flow is interrupted suddenly, the bed collapses immediately. The expansion of the bed is less pronounced, compared to group A powders (Geldart, 1973).

The largest and most dense particles are categorized in group D. These particles tend to form spouts and are ideal to be processed in spouted beds. When processed in a conventional fluidized bed, the bubbles move slower than the suspension gas ('slow bubbles'). The bed expands even less than group B particles and also collapses immediately after interrupting the gas flow (Geldart, 1973).

Particles of group C show small diameters (few µm or less) and are difficult to fluidize because of strong interparticle cohesion forces. Van der Waals forces in the size range of

group C particles are in the order of magnitude or larger than the gravitational forces and thus, dominate the behavior in fluidized beds. This results in the formation of agglomerates and channeling during fluidization, or the fluidization of Geldart C powder is not possible at all without assistance to overcome the cohesive forces (Geldart, 1973, Kunii et al., 2013).
For nano-particles it was observed, that fluidization is possible without assistance, but at gas velocities much higher than expected for the primary particles. This is due to the formation of dynamic agglomerates which constantly form, fall part and reform during fluidization. This phenomenon is also called aggregate-fluidization or agglomerate fluidization. Two types of agglomerate fluidization are distinguished. One is the agglomerate particle fluidization (APF). It is accompanied with a large expansion of the bed as soon as the gas flow is introduced into the bed. This bed expansion already occurs well before the fluidized bed state is achieved (Nam et al., 2004, Raganati et al., 2018). The other type is called agglomerate bubbling fluidization (ABF), which is characterized by a much lower bed expansion (compared to APF) and the occurrence of bubbles as soon as $u_{\mathrm{mf}}$ is reached (Gündoğdu and Tüzün, 2006, Raganati et al., 2018). However, the fluidization of nanoparticles exceeds the scope of this thesis. Thus, it is only mentioned here for the sake of completeness.

### 1.1.1 Drying in Fluidized Beds

'Drying' generally describes the removal of moisture from solids (Tsotsas and Mujumdar, 2011). Drying on industrial scale is widely applied, e.g. to reduce transportation costs by reduction of mass, increase shelf live of food and pharmaceutical products or achieve desired properties or functionalities. This thesis focuses on drying of particulate solids in fluidized beds. The drying process of a bulk of porous solids in fluidized beds, or a single porous particle in an air stream, is divided into characteristic drying phases or drying periods (DPs): a) the warm up phase, b) the first drying period (1st DP) and c) the second drying period (2nd DP) (Mujumdar and Devahastin, 2003). The respective periods are shown exemplary in Figure 1.3.
The warm up period is a short phase at the beginning of the drying process. Herein, the particle is warmed up to the wet bulb temperature $T_{\mathrm{wb}}$ and some moisture is evaporated simultaneously. Point A in Figure 1.3 marks the end of the warm up period and the beginning of the first drying period. Here, the entire particle surface is covered with water. Hence, evaporation kinetics are solely determined by gas side properties, such as temperature, flow rate and inlet moisture content of the drying gas. Therefore, the particle surface temperature as well as the temperature inside the particle equals $T_{\mathrm{wb}}$. The evaporation of water takes place at a constant rate. Thus, the first drying period is also called 'constant rate period' (Tsotsas et al., 2000).
The evaporation rate is reduced, when the particle surface is not completely covered with water. This point (B) marks the transition to the second drying period. As a consequence, the particle surface is in direct contact with the drying gas, resulting in

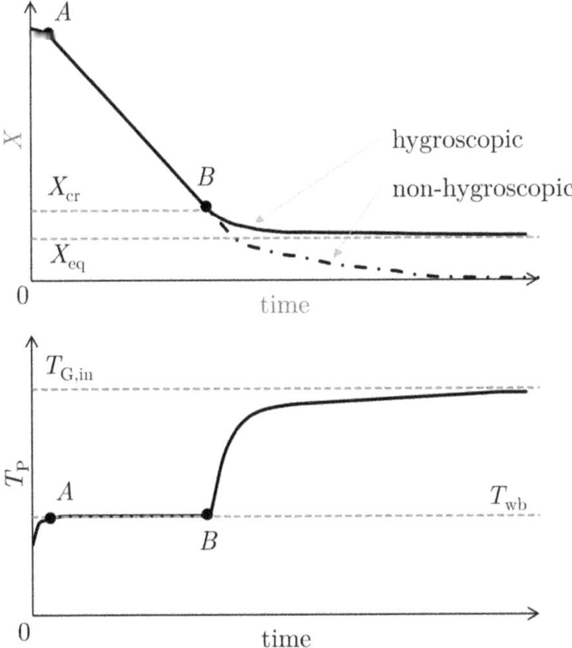


Figure 1.3: Exemplary course of drying of porous particles in fluidized beds, showing particle moisture content ($X$) and particle temperature ($T_{\mathrm{P}}$) against time, (top) and (bottom) respectively; including transition points between drying periods.

heating of the particle. In the second drying period or 'falling rate period', intraparticle transport phenomena determine the overall drying rate. These are transport of liquid water and diffusion in pores and capillaries. Simultaneous heating of the particle as well as potential changes of pore size, structure or tortuosity are superimposed and influence each other. The ultimate consequence is asymptotic approach of particle temperature towards the inlet gas temperature with passing time. Simultaneously, particle moisture content approaches the equilibrium moisture content. The equilibrium between particle moisture content and drying gas is zero for non-hygroscopic materials. The majority of porous particles are hygroscopic, meaning that the equilibrium moisture content is larger than zero, as long as the drying gas is not entirely free of moisture at the inlet (Tsotsas et al., 2000).

Due to the complexity of the involved physics in the second drying period, modeling of this period is challenging. Several approaches have been proposed and investigated. A detailed review of drying kinetics in fluidized beds and respective modeling approaches is conducted in chapter 2.2.

### 1.1.2 Assisted Fluidization

Many products, produced by fluidized bed drying, have small particle sizes (Geldart groups A and C) (Mawatari et al., 2015) or exhibit cohesive behavior due to their chemical composition (e.g. amorphous lactose) (Palzer, 2005, Piseckÿ, 2012). The presence of water during drying processes poses an additional source of potentially strong cohesive forces. As discussed above, fluidization of cohesive powders is often hindered by channeling (rat holes) and formation of agglomerates or stagnant regions in the bed. Several methods have been identified to improve fluidization in the mentioned cases. Such methods are pulsation of air flow, introduction of acoustic or mechanical vibration, magnetic or electric fields, modification of internals (i.e. baffles, agitators) or combinations of the aforementioned. The common purpose of all these approaches is counteracting cohesive forces by introduction of additional forces. Thereby, the target is to improve, or in some cases, enable fluidization, by reducing agglomeration and channeling (van Ommen, 2009).

Mechanical vibration of the fluidized bed apparatus is the option, which is most commonly applied on industrial scale for fluidized bed drying (Piseckÿ, 2012). As such vibrated fluidized beds are common in the food and pharmaceutical industry, they have been object of many investigations (Brennan et al., 2008, Gupta and Mujumdar, 1980, Jia et al., 2015, Kage et al., 1999, Kósa and Verba, 2001, Mawatari et al., 2015, 2003, Wang et al., 2000). The intended goal is to overcome interparticle forces by added vibration energy. This results in breakage of agglomerates and channels, respectively the reduction of cohesiveness. The vast majority of experimental research has been preformed in lab-scale fluidized beds with diameters ranging between a few centimeters and a couple of decimeters, or in pseudo two dimensional beds. Vibration of fluidized beds is generally reported to result in enhanced or improved fluidization. This means in particular reduction of $u_{\mathrm{mf}}$, increase of bed pressure drop and bed expansion. Furthermore, a decrease in channeling was observed with increasing vibration. Another general observation is, that the effect of vibration is stronger for Geldart group C particles than for group A and B (Mawatari et al., 2015, Xu and Zhu, 2006).

In other cases, vibration was found necessary to enable fluidization of cohesive powders in the first place (Gupta and Mujumdar, 1980, Noda et al., 1998). Furthermore, Xu and Zhu (2005) observed that segregation (of smaller agglomerates at the top and larger agglomerates at the bottom of a fluidized bed) could be reduced by vibration. Reduced segregation, respectively better mixing in VFBs, was also reported by Lee et al. (2020). The strength of vibration is often expressed as the dimensionless vibration number, which is defined as the ratio of acceleration due to vibration and gravitational acceleration:

$$\Lambda = \frac{(2\pi \cdot f)^2 \cdot A_{\mathrm{vib}}}{g}. \tag{1.1}$$

Therein, $f$ is the frequency, $A_{\text{vib}}$ the amplitude of vibration and $g$ the acceleration due to gravity. The sole characterization of vibration effects with $\Lambda$ is questionable, because different combinations of frequency and amplitude can result in the same value of $\Lambda$. It was shown, that significant differences in fluidization behavior resulted for constant $\Lambda$, but for different combinations of $f$ and $A_{\text{vib}}$. Thus, amplitude and frequency should always be considered individually (Daleffe et al., 2005, Meili et al., 2012).
Besides frequency and amplitude, the direction of vibration affects the fluidized bed (Mawatari et al., 2001). Xu and Zhu (2006) investigated different angles of vibration. For glass beads, their results show that horizontal vibration leads to lower $u_{\text{mf}}$ with increasing frequency, whereas vertical vibration results in lower $u_{\text{mf}}$ for lower frequencies. Accurate modeling of such effects is a key aspect of this thesis and an important part in the proposed flowsheet simulation model for vibrated fluidized bed dryers. The effects of vibration on hydrodynamics and drying kinetics in fluidized beds are elucidated in more detail in chapter 2.1.3.

## 1.2 Flowsheet Simulation of Solids Processes

Flowsheet simulation tools have been developed since the late 1970s. Their primary goal is the simulation of process chains, consisting of different unit operations. Key applications of flowsheet simulation are a) modeling of process behavior, b) sensitivity analysis, c) process optimization and d) process control. The vast majority of the underlying models are of empirical or semi-empirical nature. This macroscopic modeling approach allows for simulation of large time scales of complex and interconnected process chains in reasonable computational time (Gleiss et al., 2017, Puettmann et al., 2012).
The original focus lay on liquid and gas phase processes, allowing for the description of the phase properties by bulk parameters, which are often thermodynamic parameters. The detailed description of particulate materials requires consideration of interdependent multidimensional distributed parameters (Dosta et al., 2020, Skorych et al., 2020). In the example of drying, the moisture content changes. This entails simultaneous change in density and potentially changes in temperature, surface composition or size. The resulting systems of partial and ordinary integro-differential equations require special solvers and data handling algorithms (Skorych et al., 2019). Some commercial flowsheet simulation packages are nowadays available for solid processes, such as Aspen Plus (*Aspen Technology Inc.*), JKSimMet (*JKTech Pty Ltd.*), gPROMS Formulated Products (*Process Systems Enterprise Ltd.*) or CHEMCAD (*Chemstations Inc.*). An open-source alternative is the framework DYSSOL. It was developed specifically for solid processes, within the DFG priority program SPP 1679 (Heinrich, 2020, Skorych et al., 2020). It allows for steady-state and time-dynamic flowsheet simulations and is used in the scope of this work.
DYSSOL is based on the sequential-modular approach. This allows for the application of different solution approaches in different units, which is highly advantageous when

units with different multidimensional population balance models need to be solved. The waveform relaxation approach is used to increase efficiency in data transfer between units and improve convergence of the flowsheet (Skorych et al., 2020). In the DYSSOL framework, transformation matrices are used for the calculation of the multidimensional distributed parameters. Implicit calculation of transformation matrices for every unit model allows for the efficient calculation of holdups and input/output streams for the combination of process units (Skorych et al., 2019, 2017).

## 1.3 Objectives and Strategy

Drying is an important but very energy intensive process. Thus, even small improvements may lead to substantial savings in energy and costs. Fluidized bed drying is the prime choice when heat sensitive product need to be dried, which are very common in the pharmaceutical and food industry. As many products in these industries are fine and cohesive, mechanical vibration of the dryer is used to enable or improve fluidization. As fluidized bed drying is often one of many unit operations in industrial production processes, flowsheet simulation is a powerful tool for investigation and optimization of entire process chains. However, all unit operations must be modeled accurately and with reasonably low computational demand, respectively computational time. Thus, quality and robustness of the underlying model equations are paramount.
Fluidized bed drying has been investigated intensively for decades. A number of models has been proposed, of which only a few are available in flowsheet simulation frameworks. These models are sometimes only tested for a small range of process parameters and particles of only one Geldart group (Alaathar, 2017, Alaathar et al., 2020), despite significant differences and resulting deviations. Others suffer from constraints imposed by the simulation framework, e.g. that distributed material properties cannot be considered. Furthermore, the influence of vibration is not yet included in established models. Thus, the goal of this thesis is the development of a fluidized bed drying model that covers the influence of mechanical vibration of the dryer and its implementation in an open-source flowsheet simulation framework. The model is tested for a variety of particles, belonging to different Geldart groups, as well as varied dryer geometries and ranges of process conditions related to food and pharmaceutical industry. Therefore, models of fluidized bed hydrodynamics need to be developed, accounting for the impact of vibration and residence time distribution of particles, in dependence of process parameters and dryer geometry. Suitable semi-empirical correlations are identified by comprehensive experimental parameter studies.
The newly developed model parts are combined with established and reliable models and implemented in the open source flowsheet simulation framework DYSSOL. During the model development and implementation, the focus lies on the broadest possible application range of the model. Thorough validation with experimental data and sensitivity analysis are conducted to prove accuracy and reliability of model predictions, confirm

underlying assumptions and identify potential weak points as well as optimization parameters.

## 1.4 Outline of the Thesis

This thesis covers experimental investigations for the development as well as validation of a flowsheet simulation model for vibrated fluidized bed dryers. In chapter 2 modeling approaches for fluidized bed hydrodynamics, particle residence time distribution and drying kinetics are reviewed. Suitable model correlations are identified alongside areas that require further research. Furthermore, corresponding experimental investigation techniques are introduced and discussed.

Chapter 3 contains the experimental methodology of the thesis. The investigated materials are characterized with regard to properties, relevant in fluidized bed drying and the respective measurement procedures are introduced. This is followed by a detailed introduction of the different fluidized bed dryers and the installed measurement equipment, as well as specifics about the experimental procedures, used in the scope of this work.

The discussion and interpretation of experimental results is divided into two parts. Chapter 4 is concerned with the development of suitable correlations to accurately model fluidized bed hydrodynamics under the influence of vibration as well as for fine and cohesive particles. Additionally, residence time characteristics are investigated experimentally and the results are included into the model. Modeling of material specific drying kinetics of the investigated powders follows as the third pillar of the model. Lastly, the developed model and underlying assumptions are introduced in detail, the chosen correlations are discussed and the implemented structure of the different model parts is established.

The model performance is thoroughly explored in chapter 5. Model predictions are validated by comparison to experimental data. A wide range of process parameters and particles of different Geldart groups is tested, as well as the influence of different dryer geometries and vibration on the drying process and model predictions. Subsequently, crucial assumptions and model stability are examined. Additionally, sensitivity analysis with respect to process parameters is conducted. Potential fitting parameters are identified and the overall model performance is evaluated.

Concluding remarks are made in chapter 6. Here, the thesis is summarized, key findings are highlighted and topics for further research and options for continued development of this fluidized bed drying model are pointed out.

# 2

# Modeling of Fluidized Beds

## 2.1 Modeling of Fluidized Bed Hydrodynamics

The minimum fluidization velocity $u_{\mathrm{mf}}$ is commonly used as a key parameter in the design and the modeling of fluidized bed systems in the bubbling regime. Hence, it is of great interest to this thesis. Determination of $u_{\mathrm{mf}}$ is done by measuring the pressure drop across the particle bed at varied gas flow rate. The pressure drop in the fixed bed range increases with increasing gas velocity. Laminar flow through the bed entails linear increase in pressure drop, while non-linear increase is the result of turbulent flow. In the fluidized state, the pressure drop across the bed is constant. The measured pressure drop is plotted against the gas velocity and the two pressure drop lines are extrapolated. The intersection of the lines corresponds to $u_{\mathrm{mf}}$ (Kunii et al., 2013). This is shown exemplary in Figure 2.1.
The complete fluidization velocity $u_{\mathrm{cf}}$ on the other hand describes the lowest gas velocity, at which all particles are fully fluidized. Hence, it defines the lowest possible operation gas velocity of fluidized beds. As illustrated in Figure 2.1, $u_{\mathrm{cf}}$ and $u_{\mathrm{mf}}$ may vary significantly, especially for fluidized beds with wide particle size distributions. Large particles require higher gas velocities to be fluidized than smaller particles of equal density. Furthermore, cohesive forces may result in channeling in the fixed bed regime, rendering the traditional method of determination of $u_{\mathrm{mf}}$ impossible. In such cases, $u_{\mathrm{cf}}$ is an alternative to characterize the onset of fluidization by pressure drop measurements (Kono et al., 1986, Liu et al., 2016).

### 2.1.1 Modeling Approaches

As mentioned above, the bubble characteristics in fluidized beds depend on the Geldart group of the fluidized particles. Furthermore, the bubble size depends on the type of distributor, excess gas velocity (difference of superficial gas velocity and minimum fluidization velocity $u_0$–$u_{\mathrm{mf}}$) and the height above the distributor. Experimental data indicates that bubble size and bubble volume fraction (or gas hold-up) increase with increasing superficial gas velocity as well as increasing bed height (Karimipour and Pugsley, 2011).

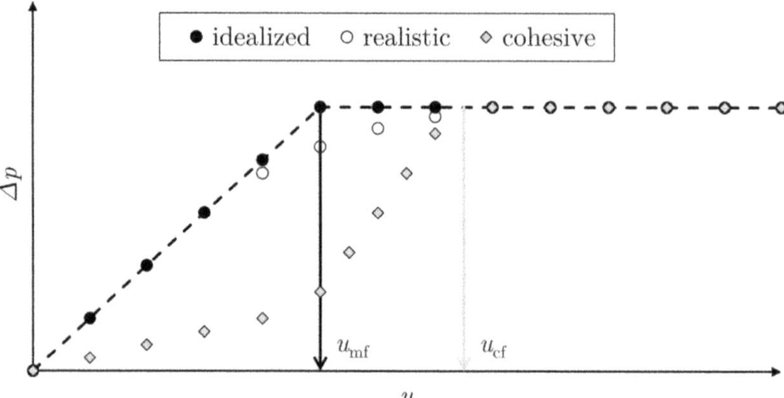


Figure 2.1: Pressure drop across bed material against gas velocity and determination of minimum fluidization velocity ($u_{\text{mf}}$) and complete fluidization velocity ($u_{\text{cf}}$) for idealized (mono-sized, spherical particles), realistic (broader size distribution) and cohesive (with channeling in the fixed bed) bed material; adopted from Lehmann et al. (2019).

The vessel dimensions are not only the limit of the bubble diameter, the dimensions also influence the hydrodynamics (Werther, 1974). According to the original two-phase theory (Davidson and Harrison, 1966), all excess gas ($u_0$–$u_{\text{mf}}$) passes through the bed in the form of bubbles. However, this has been disproved by many researchers (Hilligardt and Werther, 1986, Karimipour and Pugsley, 2011, Zou et al., 2011). Hilligardt and Werther (1986) showed that the factor by which the excess gas velocity deviates from the actual visible bubble flow $\dot{V}_{\text{B}}$ strongly depends on the bed dimensions. Furthermore, they showed that the bubble rise velocity $u_{\text{B}}$ also depends on the bed diameter. The particle properties, classified according to Geldart groups, also play a significant role in the bubble characteristics (Hilligardt and Werther, 1986).

Numerous correlations for bubble size and bubble rise velocity have been developed over the decades. Karimipour and Pugsley (2011) reviewed 25 correlations for bubble diameters and seven correlations for bubble rise velocity. The bubble size is commonly described by an equivalent bubble diameter. These correlations have been developed for different types of particles (Geldart groups A, B and D) as well as under different operation conditions, such as superficial gas velocity or bed dimensions. Consequently, the correlations vary significantly in their predicted results and are limited in their application range. The most commonly applied model for the bubble characteristics was proposed by Hilligardt and Werther (1986). They validated their model for Geldart A, B and D particles and a wide range of bed geometries. More recent investigations (Bakshi et al., 2017, Haghgoo et al., 2018) affirmed high accuracy of the predictions of the Hilligardt and Werther (1986) model.

The advantage of the Hilligardt and Werther (1986) model over the other models reviewed in Karimipour and Pugsley (2011) is that it allows for calculation of the bubble size, and also estimation of the bubble volume fraction. The bubble volume fraction is the key parameter describing the hydrodynamics in the simulation of bubbling fluidized beds. Thus, the Hilligardt and Werther (1986) model has been widely applied in the simulation of fluidized beds in drying (Alaathar et al., 2013, Burgschweiger and Tsotsas, 2002, Groenewold and Tsotsas, 2007), chemical looping combustion (where the bubbling part of the bed is referred to as dense bottom zone) (Haus et al., 2018, Kramp et al., 2012) and catalytical co-polymerization (Abbasi et al., 2019).

The velocity of the bubbles, often called bubble rise velocity, is directly related to the size of the bubbles (Karimipour and Pugsley, 2011). However, several different correlations have been proposed. They mostly differ in the constants used in the semi-empirical equations, resulting from different measurement techniques, particles and size of the fluidized bed used in the respective experiments or numerical investigations. Seven correlations for the bubble rise velocity were reviewed by Karimipour and Pugsley (2011). More recently, Agu et al. (2019), Maurer et al. (2016), Puncochar et al. (2016), Wei et al. (2017) proposed own models for the bubble size and rise velocity. Introduction and discussion of all the existing models would exceed the scope of this review.

Next to the bubble characteristics also the resulting bed expansion as well as the ratio of bubble and suspension phase (also referred to as bubble volume fraction or bubble hold-up) are crucial for a complete model of the hydrodynamics of fluidized beds. The expansion of the bed is the increase of porosity when the bed is fluidized compared to the fixed bed state. The porosity of the bed during fluidization is commonly estimated with an equation proposed by Richardson and Zaki (1954). Therein, the bed porosity $\varepsilon$ is expressed using the ratio of the superficial fluid velocity $u$ and the terminal velocity $u_\mathrm{t}$ of a particle in said fluid as well as an empirical exponent $n$:

$$\varepsilon = 1 - \left(\frac{u}{u_\mathrm{t}}\right)^{1/n} . \tag{2.1}$$

In the original version of particles in liquid, Richardson and Zaki (1954) identified $n$ to be 4.65. This value for the Richardson Zaki exponent is still used today to estimate the expansion or bed porosity of gas fluidized beds. However, several researchers published modified versions of equation 2.1 over the years. Mainly, the value of the Richardson Zaki exponent $n$ was varied or calculation methods for it were proposed (Martin, 2010). Further modifications of the Richardson Zaki equation or simply the exponent have been proposed to account for the temperature influence (Lettieri et al., 2002), the influence of magnetic fields on the fluidization of magnetic particles (Valverde and Castellanos, 2008) as well as the influence of non-spherical and multi-sized particle systems (Bargieł and Tory, 2013), to only name a few.

However, the vast majority of the mentioned empirical or semi-empirical models are based on experimental data. Thus, every model is only as good as the data it is based on and validated against. Some of the above mentioned models are based on numerical simulations, which need to be validated with experimental data beforehand. Experimental techniques, suitable for investigations of fluidized bed hydrodynamics, are introduced and reviewed with regard to applicability to this thesis in the following chapter.

### 2.1.2 Experimental Techniques

Information about the flow regime and thus, the hydrodynamics in fluidized beds can be gathered via measurement of pressure signals. This is particularly interesting for industrial applications, because industrial fluidized beds are commonly made of steel and have little or no windows for visual observations. Furthermore, pressure measurements are comparably cheap, robust and operational in harsh conditions, i.e. high temperatures. The pressure signal is influenced by different phenomena in the fluidized bed. The passing of a bubble along the pressure measurement port results in a different pressure reading compared to particles (suspension phase) being present at the port due to the pressure difference between bubbles and the suspension. Besides these local phenomena, global effects are detectable. Global effects influence the pressure signal by sending pressure waves through the system. These are the eruption of bubbles at the surface and vibrations in the system. There may be intended vibration or hammering of the apparatus but also unintended ones, e.g. vibrations in the periphery, such as rotational speed of the blower, or fluctuations of the gas flow (Bi, 2007, van Ommen et al., 2004). In order to capture said effects, pressure signals should be recorded at high frequencies (>50 Hz). The pressure signal may be analyzed in the time domain and in the frequency domain. In the time domain analysis, the standard deviation of the pressure signal is plotted against time. Certain effects have been attributed to the absolute value of the standard deviation of the pressure signal. For instance, the transition from bubbling to turbulent fluidization is characterized by an increase in standard deviation in the transition region, whereas in the bubbling and in the turbulent regime the standard deviation is of similar magnitude. By plotting averaged values of the standard deviation of the pressure signal, taken at different gas velocities, the transition velocity from bubbling to turbulent fluidization may be identified (Bi et al., 2000).
For the analysis of the pressure signal in the frequency domain, Fourier transform is applied to the pressure signal, leading to the power plot of the pressure signal. Every fluidized bed has a dominant frequency, which is represented by a peak in the power plot. The value of said frequency depends on particle properties, gas velocity and apparatus dimensions. It usually lies in a range from 2 Hz to 15 Hz. Depending on the flow regime, the peak at the dominant frequency is sharp or the pressure signal is scattered around the dominant frequency. The narrower the pressure signal around the peak, the more stable or ordered is the flow regime. A large scatter indicates unstable flow regimes or transitions in the flow regime (Felipe and Rocha, 2004).
Felipe and Rocha (2004) identified the windbox of fluidized beds as an ideal position for the pressure probe to analyze the global pressure signal in terms of the flow regime of the fluidized bed. Positioning the probe inside the bed will automatically lead to a certain

overlap of local bubble effects and global signals (Felipe and Rocha, 2004, van Ommen et al., 2004). Thus, more care must be taken during the analysis of pressure signal from inside the bed. On the other hand, pressure signals from inside the bed were proven to be applicable for the measurement of local bubble characteristics (Zhang et al., 2020).

A number of methods for experimental investigation of bubble characteristics, such as diameter and velocity, have been applied in the literature. They can be categorized into intrusive and non-intrusive techniques and have been reviewed briefly by Karimipour and Pugsley (2011). Intrusive measurement techniques such as optical, capacitance, electro resistivity or conductivity probes are located inside the bed and will thereby influence the hydrodynamics and impair the data (Karimipour and Pugsley, 2011). However, only intrusive probes are applicable for the measurement of bubble characteristics inside large pilot-plant-scale and industrial fluidized beds.

Optical probes have a light emitting and a light receiving part installed in the tip of the probe. The amount of reflected light is measured and correlated to the presence of particles (more light is reflected) and bubbles (less reflection). The placement of two probes at a known difference above each other allows for estimation of the bubble rise velocity by comparing both signals via cross-correlation (Rüdisüli et al., 2012).

Capacitance, electro resistivity and conductivity probes use the difference in electric-physical properties (dielectric constant, electrical resistance or conductivity, respectively) between the gas and the particles. Thus, passing bubbles result in a different voltage signal compared to particles being present between two electrodes (Park et al., 1969, Werther and Molerus, 1973). Estimation of bubble velocities is done in the same way as for optical probes (Wiesendorf and Werther, 2000). Cylindrical capacitance probes were introduced by Werther and Molerus (1973) and were used most commonly in various types of fluidized beds (Karimipour and Pugsley, 2011, Wiesendorf, 2000). Werther and Molerus (1973) showed that small, cylindrical capacitance probes have very little influence on the hydrodynamics of the bed. They observed that the signal of a single probe mounted in the bed showed the same power spectral density as the same probe but with an identical probe mounted beneath it.

However, all of the described intrusive measurements can only give insight into the local bubble characteristics. In order to get global information, external methods are required. Such non-intrusive techniques are direct (video) photography through transparent apparatus walls or of the bed surface. They only yield information about the observed surface. Thus, they are commonly applied for investigations in 2D fluidized beds. However, bubble behavior at the apparatus wall may differ significantly from the inside of the bed. Significant differences in bubble behavior between 2D and 3D beds have been reported (Hilligardt, 1986).

An alternative method for the investigation of bubble velocities in fluidized beds is the tracer gas method. Thereby, a pulse of inert tracer gas is introduced into the bed. At a position further up in the bed, the concentration of that tracer gas is measured. By correlating the time lag and the known distance between source and measurement point of the tracer gas, the rise velocity of the gas is measured. It is assumed that the tracer gas travels in bubbles though the bed. However, the exact path of the tracer gas is unknown. It may travel in an individual bubble, multiple bubbles and the suspension gas. Comparison of tracer gas measurements to other measurement principles suggested the tracer gas method's applicability to measure bubble velocity (Dry et al., 1984). Furthermore, the dispersion of the tracer gas over the bed height was measured with the tracer gas method (Zhang et al., 2018). However, no information is attainable if the tracer is present in bubbles or suspension. Consequently, the tracer method on its own is not applicable for a detailed investigation of bubble characteristics such as size, shape and velocity distributions.

Other non-intrusive techniques, such as X-ray or electrical capacitance photography, have been used to gather valuable data on bubble characteristics in 3D beds (Fabich et al., 2017, Maurer et al., 2016, Penn et al., 2020). However, their key disadvantages are complex and costly equipment, and limitation to small, lab-scale fluidized beds of only a few cm in diameter.

### 2.1.3 Vibrated Fluidized Beds

The influence of vibration on bubble characteristics in 2D fluidized beds has been the focus of some researchers (Cano-Pleite et al., 2014, Eccles and Mujumdar, 1997, Zhou et al., 2004) and their respective groups. Cano-Pleite et al. (2016) summarized those findings and pointed out that seemingly contradicting results are due to differences in experimental settings and evaluation. The main reason was that bubble characteristics were only evaluated in small fractions of the bed (stripes no larger than 6 cm in height). By evaluating the local bubble characteristics as a function of bed height with digital image analysis, Cano-Pleite et al. (2016) found that bubbles in the lower part of the bed behave very similarly to isolated bubbles, whereas in the upper parts of the bed, coalescence dominates the bubble behavior. Another observation was that with increased vibration, bubbles tend to move preferably to the middle of the bed and form a core-annulus flow (or 'gulf stream circulation'). This confinement of the bubble pathway enhances coalescence of bubbles. Taking the average over the entire bed, a decrease of bubble volume fraction was observed with increasing vibration intensity (constant frequency of 20 Hz and increasing amplitude from 0 mm to 1.9 mm). Particularly for fine particle beds, when vibration is introduced into the system, the coalescence of bubbles is promoted. Also, the visible bubble flow $\dot{V}_B$ is reduced by suspension gas, taking shortcuts by flowing through the bubbles. Additionally, increasing vibration was observed to increase bubble size (Cano-Pleite et al., 2016).

The bed (particles and the gas, not the apparatus) are an under-damped vibrating system and therefore have a resonance frequency. Hence, vibrating a fluidized bed at the resonance frequency has stronger, or even different effects, compared to other vibration frequencies. Studies by Xu and Zhu (2005) showed that initially the size of agglomerates inside a VFB decreased with increasing vibration intensity. At a certain vibration intensity, the agglomerate size began to increase. Next to the agglomerate size, also the flow regime of the fluidized bed was reported to change significantly at a critical vibration intensity (Valverde and Castellanos, 2006). This phenomenon can be explained by finite size reduction with increasing vibration energy. Additionally, the intensity of the agglomerate contact increases, stimulating agglomerate growth with increasing vibration intensity. This critical vibration intensity was found to correlate to the natural or resonance frequency of the bed, whereas the amplitude has no significant impact (Barletta and Poletto, 2012, Valverde and Castellanos, 2006, Xu and Zhu, 2005).

Wang et al. (2002) investigated VFBs with regard to their resonance frequency and showed that bed expansion increases until the resonance frequency is reached. With further increase, the bed expansion stayed constant. However, caution is advised here, because the experimental setup did not control the amplitude of vibration, which is thereby a result of the vibration frequency. Hence, the resonance frequency of the vibration table (consisting of the spring suspension) was observed (Wang et al., 2002, Wensrich and Collard, 2006). This could have led to overlapping effects (Wensrich and Collard, 2006).

Eccles and Mujumdar (1992) reported significant influence of the resonance frequency of the bed. They investigated the heat transfer from a VFB to a cylindrical probe inside the bed. They also made the same observations regarding flow characteristics in dependence of the resonance frequency as Barletta and Poletto (2012), Valverde and Castellanos (2006), Xu and Zhu (2005). Additionally, Eccles and Mujumdar (1992) observed increasing heat transfer rates with increasing vibration frequency, reaching a maximum at resonance frequency and slightly decreasing with further increase in vibration frequency. Thus, knowledge of the resonance frequency of the bed is another important parameter in the modeling of the hydrodynamics of VFBs.

Bi (2007) reviewed several correlations for characteristic frequencies in fluidized beds. The correlation from Gidaspow et al. (2001) is found to reflect the resonance frequency of the bed material investigated in the scope of this thesis. It is described by:

$$f_{\mathrm{res}} = \frac{1}{2\pi}\sqrt{\frac{g}{H_{\mathrm{mf}}}\frac{\left(3\left(\varepsilon^{-1}-1\right)+2\right)\left(1-\varepsilon\right)}{1-\varepsilon_{\mathrm{mf}}}}\,. \tag{2.2}$$

Despite the intensive research on vibrated fluidized beds (see above cited literature) no comprehensive model for the hydrodynamics in VFBs has been published. Merely a few models to predict the bed expansion have been proposed in literature. Jin et al. (2007) proposed a model for Geldart B and D particles in VFBs. However, the dominance of the cohesive forces increases with smaller particle size. Thus, the impact of vibration is stronger for particles of Geldart groups A and C, which is likely to require different correlations. Hence, the development of a model to predict fluidized bed hydrodynamics of fine and cohesive powders is a key part of this thesis.

### 2.1.4 Residence Time Distribution

Fluidized bed processes are operated in batch and continuous mode. In batch operations, the bed material is filled before the process is started and the entire bed is emptied after the process is finished. All particles stay in the system for the same period of time. Under continuous operation, feed particles are constantly conveyed into the fluidized bed and leave the process after a certain time in the fluidized bed. The residence time describes the amount of time, a particle stays in a continuous fluidized bed, i.e. the time it needs to travel from inlet to outlet of an apparatus (Levenspiel, 1999). Particles, entering an apparatus at the same time are likely to show different residence times due to different paths and trajectories inside the apparatus. This results in a distribution of residence time in the process. The residence time distribution (RTD) has significance influence on the process and product quality. Two theoretical border cases for RTDs have been defined: the plug flow reactor (PFR) and the ideally mixed continuous stirred tank reactor (CSTR) (Levenspiel, 2014).
In an ideal PFR, all particles have the same residence time, i.e. travel through the process without deviating paths from each other. In contrast, the CSTR is characterized by ideal mixing, meaning that particles entering the process are immediately mixed with the particles already present, resulting in homogeneous, location-independent concentration in the reactor. The actual RTD in fluidized beds lies between the two ideal cases. In general, the RTD of long and narrow, horizontal fluidized beds is closer to PFR characteristics, whereas cylindrical or vertical fluidized beds with aspect ratios near unity will show RTDs similar to the CSTR (Levenspiel, 2014).
RTDs are quantified by the exit age distribution. Steady-state conditions and incompressible flow are assumed, as well as that transport in and out of the process is governed by forced convection. Additionally, it is assumed that particles cannot leave the process through the inlet and cannot re-enter through the exit, thereby fulfilling the closed-closed boundary condition, defined by Danckwerts (1953). The exit age $E$ of the particles, introduced to the system at the same time and leaving the process in a defined time interval, is plotted against time $t$. By definition, the area under this exit age curve $E(t)$ is unity.

$$\int_0^\infty E(t)dt = 1, \tag{2.3}$$

representing the density distribution of the RTD. The function $F(t)$ sums up the ratio of particles with exit age lower than $t$. This cumulative distribution is defined as:

$$F(t) = \int_0^t E(t)dt. \tag{2.4}$$

The mean residence time $\tilde{\tau}$ is defined as the first moment $\mu_1$ of the exit age distribution or E-function in short:

$$\mu_1 = \tilde{\tau} = \int_0^\infty t^1 \cdot E(t)dt. \tag{2.5}$$

Given that there are no dead zones in the reactor, no bypass streams nor changes in density, the mean residence time equals the hydrodynamic residence time $\tilde{\tau}$, which is defined as the ratio of particle mass hold up and particle mass flow:

$$\bar{\tau} = \frac{\dot{M}_{\mathrm{P}}}{M_{\mathrm{P}}} \tag{2.6}$$

The second moment of the E-function $\mu_2$ is used to determine the variance $\sigma^2$, representing the width of the RTD:

$$\mu_2 = \int_0^\infty t^2 \cdot E(t)dt, \tag{2.7}$$

$$\sigma^2 = \int_0^\infty (t - \tilde{\tau})^2 \cdot E(t)dt = \mu_2 - \mu_1^2 . \tag{2.8}$$

In an ideal PFR, $\sigma^2 = 1$; an ideal CSTR is characterized by $\sigma^2 = 0$ (Nauman, 2008). In order to simplify comparability, dimensionless description of RTD is advantageous. Therefore, the dimensionless residence time $\theta$ is defined:

$$\theta = \frac{t}{\tilde{\tau}} . \tag{2.9}$$

Dimensionless expression of the density and cumulative distributions are given by:

$$E(\theta) = \tilde{\tau} \cdot E(t), \tag{2.10}$$

$$F(\theta) = F(t) . \tag{2.11}$$

For the description of RTD characteristics of real processes, two model approaches are commonly used: 1) the cascade or tank in series (TIS) model and 2) the dispersion model. Both are introduced in the following sections. Experimental approaches and findings with respect ot fluidized bed processes are reviewed subsequently.

### Tank in Series Model

The cascade or TIS model, introduced by MacMullin and Weber (1935), assumes that the RTD of a given process can be described by a series of identical CSTRs. The density distribution of residence time is expressed as:

$$E(t) = \frac{1}{\tilde{\tau}} \left(\frac{t}{\tilde{\tau}}\right)^{K-1} \frac{K^K}{(1-K)!} \exp\left(-\frac{t}{\tilde{\tau}} K\right), \tag{2.12}$$

where the theoretical number of tanks $K$ is defined as:

$$K = \frac{\tilde{\tau}^2}{\sigma^2} . \tag{2.13}$$

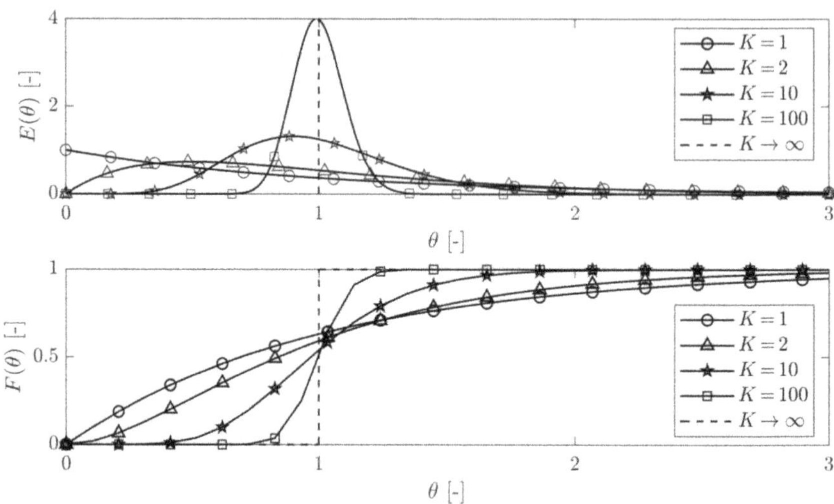


Figure 2.2: Examples of residence time distribution curves after the tank-in-series (TIS) model, for different number of tanks, plotted against dimensionless residence time ($\theta$), showing corresponding density distributions (top) and cumulative distributions (bottom).

In the dimensionless notation, the E-function of the RTD is denoted as:

$$E(\theta) = \frac{K\,(K\theta)^{K-1}}{(1-K)!}\exp\left(-K\theta\right). \tag{2.14}$$

CSTR behavior is represented when $K=1$. If $K$ approaches infinity, the residence time corresponds to the ideal PFR. The trends for different numbers of tanks are exemplary depicted in Figure 2.2. Chen et al. (2017) showed that the RTD in horizontal fluidized bed dryers may be described best, using equation 2.12 and positive non-integer values as theoretical tank number $K$. In this case, application of the gamma function $\Gamma(K)$ allows for accurate RTD modeling in continuous fluidized beds. The gamma function is defined as the following improper integral, valid for complex numbers with positive real parts (Davis, 1959):

$$(1-K)! = \Gamma(K) = \int_0^\infty x^{K-1}e^{-x}dx\,. \tag{2.15}$$

### Dispersion Model

A different approach, fulfilling the same goal and criteria as the TIS model, is the dispersion model. Originally introduced by Langmuir (1908) and Danckwerts (1953), it applies the analogy of molecular diffusion on particle mixing in fluidized bed systems. Considering horizontal reactors, dispersion along the horizontal axis (axial dispersion) is expressed by the axial dispersion coefficient $D_{\text{ax}}$. It includes molecular diffusion as well as dispersion of particles due to different trajectories in the reactor. The RTD may be expressed as:

$$\frac{\partial c_{\text{T}}(t)}{\partial t} = D_{\text{ax}} \frac{\partial^2 c_{\text{T}}(t)}{\partial x^2} - u_{\text{ax}} \frac{\partial c_{\text{T}}(t)}{\partial x}, \tag{2.16}$$

with $x$ denoting the horizontal coordinate. The axial velocity of the particles $u_{\text{ax}}$ is the ratio of horizontal reactor length $L_{\text{ax}}$ and mean residence time.

$$u_{\text{ax}} = \frac{L_{\text{ax}}}{\tilde{\tau}} \tag{2.17}$$

The dispersion coefficient $D_{\text{ax}}$ or the dimensionless dispersion number $D_{\text{ax}}/u_{\text{ax}} L_{\text{ax}}$ are used for characterization of RTD behavior (Levenspiel, 2014). It is similar to dimensionless numbers, describing only molecular diffusion, namely Bodenstein $Bo$ or Peclet number $Pe$:

$$Bo = Pe = \frac{u\,L}{D}, \tag{2.18}$$

where $D$ is the diffusion coefficient, $L$ the characteristic length and $u$ the flow velocity. Levenspiel (2014) pointed out, that $Bo$, $Pe$ and the dimensionless dispersion number have been used equivalently in the past, although technically not correct. In the light of this, the Bodenstein number for fluidized beds $Bo_{\text{fb}}$ is defined and used for characterization of particle residence time in the scope of this work. It is the reciprocal of the dimensionless dispersion number and accounts for all factors causing particle mixing in the vessel, such as molecular diffusion as well as mixing by turbulent effects, laminar flow profiles, etc.

$$Bo_{\text{fb}} = \frac{u_{\text{ax}} L_{\text{fb}}}{D_{\text{ax}}}. \tag{2.19}$$

Now, a dimensionless expression of particle dispersion is possible, using the dimensionless length $\chi = x/L_{\text{fb}}$:

$$\frac{\partial c_{\text{T}}}{\partial \theta} = \frac{1}{Bo_{\text{fb}}} \frac{\partial^2 c_{\text{T}}}{\partial \chi^2} - \frac{\partial c_{\text{T}}}{\partial \chi}. \tag{2.20}$$

The border case of ideal PFR is reached when particles show no dispersion, meaning that $D_{\text{ax}} = 0 \Rightarrow Bo_{\text{fb}} \rightarrow \infty$. Perfect mixing of the CSTR is represented by $D_{\text{ax}} \rightarrow \infty \Rightarrow Bo_{\text{fb}} \rightarrow 0$.

### Experimental Investigation of RTD

Residence time characteristics of fluidized beds are commonly investigated with tracer experiments. A known mass or volume ($M_{\text{T},0}$ or $V_{\text{T},0}$ respectively) of tracer is inserted into the feed of the fluidized bed and the concentration of tracer $c_{\text{T}}$, or a signal, proportional to the tracer concentration $S_{\text{T}}$, is measured in the outlet flow. Hence, the density distribution curve is determined by:

$$E(t) = \frac{\dot{M}_{\mathrm{T}}(t)}{M_{\mathrm{T},0}} = \frac{\dot{V} c_{\mathrm{T}}(t)}{\int_0^\infty \dot{V} c_{\mathrm{T}}(t) dt} = \frac{c_{\mathrm{T}}(t)}{\int_0^\infty c_{\mathrm{T}}(t) dt} = \frac{S_{\mathrm{T}}(t)}{\int_0^\infty S_{\mathrm{T}}(t) dt}. \tag{2.21}$$

A variety of tracers has been used to characterize RTD in solid processes. It is key to distinguish the tracer clearly from the regular bed material, while interfering with the process as little as possible. Therefore, tracer particles must fulfill a number of requirements (Gao et al., 2012, Pietsch, 2018):

- same chemical and physical properties (especially size and density in fluidized beds) and thus,
- mix well with the regular particles and not interfere with fluidization characteristics
- inert, i.e. must not adsorb substances during the process, or undergo irreversible reactions
- be detectable in small amounts

Distinction between tracer and regular solids may be based on optical properties, such as color or phosphorescence (Bachmann et al., 2016, Brod et al., 2004, Diez et al., 2019, Finzer et al., 2007, Han et al., 1991, Satija and Zucker, 1986). Also, radioactive properties (Hull and Rosenberg, 1960), magnetic or conductive properties (Idakiev and Mörl, 2013, Pietsch et al., 2018), pH value (Jacob, 2010) or solubility (Nilsson and Wimmerstedt, 1988) were used for investigations in fluidized bed processes.
The introduced exit age or RTD curves represent the systems response to a small disturbance by tracer injection. Tracer particles may be injected as pulse, step or in periodic or stochastic manner. Most common is the pulse injection, where the entire volume of tracer particles is introduced at once, practically meaning as fast and as complete as possible but not longer than one percent of $\tilde{\tau}$. Ideally, the tracer impulse is described as a Dirac impulse (Nauman, 2008):

$$\int_{-\infty}^{\infty} \delta dt = 1. \tag{2.22}$$

A number of RTD investigations in fluidized beds have been conducted in circulating fluidized beds (CFBs). These were reviewed by Gao et al. (2012) and are not further discussed here, because the RTD characteristics in CFBs differ significantly from bubbling fluidized beds and horizontal fluidized beds, relevant to this thesis.
In relevant studies, general dependencies between process parameters and RTD have been reported to be increasing dispersion coefficient or decreasing $Bo_{\mathrm{fb}}$ respectively for:

- higher gas velocity (Bachmann et al., 2016, Jacob, 2010, Nilsson and Wimmerstedt, 1988, Reay, 1978)
- lower gas temperature, respectively gas density (Nilsson and Wimmerstedt, 1988)

- increasing bed height, respectively outlet weir height (Bachmann et al., 2016, Idakiev and Mörl, 2013, Nilsson and Wimmerstedt, 1988, Reay, 1978)
- decreasing particle size and density (Bachmann et al., 2016, Nilsson and Wimmerstedt, 1988, Reay, 1978)
- increasing particle sphericity (Vollmari and Kruggel-Emden, 2018)
- decreasing particle moisture content, respectively stickiness (Khanali et al., 2012)
- lower particle mass flow (Bachmann et al., 2016, Nilsson and Wimmerstedt, 1988, Reay, 1978)

The above mentioned studies were conducted with particles of Geldart groups B and D. Furthermore, it was observed that the aspect ratio of the dryer (length over width) has significant impact on the RTD. High aspect ratios, i.e. long and narrow dryers, show less dispersion of particles and are therefore closer to the ideal PFR. For small aspect ratios, i.e. closer to quadratic geometry, the RTD characteristics approach the CSTR (Bachmann et al., 2016).

Bachmann et al. (2016) proposed a comprehensive correlation to predict the RTD characteristics via calculating the Bodenstein number $Bo_{\text{fb}}$. Their correlation accounts for bed geometry, particle properties, fluidization characteristics and process conditions; it represents own and literature data well (Bachmann et al., 2016, Nilsson and Wimmerstedt, 1988, Reay, 1978).

Another aspect of fluidized configuration that has huge impact on RTD characteristics is the installation of baffles or weirs. As weirs hinder the backmixing of particles, they lead to reduced dispersion. The more weirs installed, the lower the dispersion and approximation of ideal PFR behavior. Reduction of the distance between weirs was found to increase mean residence time (Diez et al., 2019, Satija and Zucker, 1986, Vollmari and Kruggel-Emden, 2018). Bachmann et al. (2017) proposed correlations for the Bodenstein number for overflow weirs and underflow weirs, based on their previous correlation. Since such compartment-fluidized beds are not considered in this thesis, this topic is not further pursued here.

However, the influence of vibration is an important aspect of this work. The impact of vibration on the RTD characteristics in vibrated fluidized bed dryers was investigated by only a few researchers (Brod et al., 2004, Han et al., 1991, Satija and Zucker, 1986). These studies have in common, that vibration amplitude was varied, whereas vibration frequency was kept constant in all cases. Increasing vibration was unisonously reported to reduce mean residence time, as horizontal components of the vibration movement support the transport of particles towards the outlet, thus increasing axial particle velocity $u_{\text{ax}}$. According to equation 2.18, higher dispersion number and lower Bodenstein number are the consequence. This was observed in literature studies (Han et al., 1991, Satija and Zucker, 1986).

Finzer et al. (2007) investigated the drying of coffee cherries in a vibrated tray dryer.

Although their setup varied significantly from fluidized bed dryers, their observations of increased dispersion with decreasing stickiness and increasing sphericity are analogous to observations in fluidized beds (Khanali et al., 2012, Vollmari and Kruggel-Emden, 2018). In the vibrated tray dryer, the increased dispersion was accompanied by a reduced axial particle velocity. Hence, the resulting Bodenstein numbers were almost constant in the investigated range of particle sphericty and stickiness (Finzer et al., 2007).
Due to the scarce research on the impact of vibration on RTD characteristics, this aspect will receive special attention in chapter 3.3.6.

## 2.2 Modeling of Fluidized Bed Drying Processes

Modeling of the first drying period in fluidized beds is straight forward, because the kinetics are solely influenced by gas side properties. Thus, application of Fick's law allows for prediction of drying kinetics.
However, when intraparticle resistances and transport phenomena become the rate limiting factors, the second drying period is reached and modeling becomes more complex. Kemp and Oakley (2002) reviewed modeling of fluidized bed drying and reported two categories of models for the kinetics in the second drying period: 1) the distributed parameter model (DPM) and 2) the lumped parameter model (LPM). The idea behind the DPM is to describe fundamental physical effects. However, description of fundamental physical effects by non-empirical equations is often not feasible. Crucial material parameters, e.g. diffusion coefficients, pore size, pore size distribution and tortuosity, are difficult to measure. Due to this complexity and vast experimental effort, only a very limited number of materials have been characterized to allow modeling with the DPM (Kemp and Oakley, 2002). Undoubtedly, the DMP would deliver the highest possible accuracy. Even so, highly detailed drying models may not be required. Kemp and Oakley (2002) reported that a higher level of detail is not automatically accompanied by higher accuracy of model predictions, regarding the overall drying process. Foremost, for macro-scale flowsheet simulation it is important to keep the computational effort low and, thereby, the calculation time reasonably low. Therefore, it is crucial to capture the main effects in a reliable manner, while micro-scale phenomena may be treated implicitly.
For this purpose, LPMs are designed to capture the drying kinetics by representing the involved physical effects in a few equations (sometimes just one). Hence, the drying kinetics are measured and fitted either empirically or semi-empirically. This empirical nature of the model limits the applicability to process conditions similar to the ones investigated in the model development. Numerous LPMs with different degrees of complexity have been proposed in literature. More detailed versions are shrinking core models (Defraeye et al., 2012, Hashimoto et al., 2003) which consider the receding evaporation front to the center of the particle. Along with this, additional heat and mass transfer resistance is represented by the already dry layer at the particle surface.

Among others, Mezhericher et al. (2010) used the shrinking core approach to model single droplet drying after particle formation.

Another approach has emerged in recent years. The pore network modeling has been used to elucidate physical phenomena involved in drying of porous media (Kharaghani et al., 2012, Laurindo and Prat, 1998, Surasani et al., 2008). With it, the gap towards the DPM could be narrowed, but is not yet closed (Attari Moghaddam et al., 2018, Mortier et al., 2011). Furthermore, pore network models were used to derive an overall or effective moisture transport coefficient (Jabbari et al., 2019) in the classical sense of the LPM.

The most common application of LPMs is the representation of the second drying period by one mathematical function, which often is of exponential or polynomial nature. This function is derived by non-linear regression of experimental drying kinetics data, measured in either single droplet, thin layer or batch fluidized bed drying experiments. The regression function may be expressed as the dimensionless moisture ratio ($MR$) (Kemp and Oakley, 2002):

$$MR = \frac{X(t) - X_{\mathrm{eq}}}{X_0 - X_{\mathrm{eq}}} = f(t), \tag{2.23}$$

where $X_{\mathrm{eq}}$ is the equilibrium moisture content and $X_0$ the initial moisture content. Using this function, or drying curve, drying kinetics under different process conditions are estimated. Various mathematical forms have been proposed in literature for the drying curve (Chen and Putranto, 2013, Ertekin and Firat, 2017). The optimal shape and fit of the function depends on the investigated material and drying conditions. Ertekin and Firat (2017) provided a comprehensive overview of types of empirical drying curves. The approach of empirical drying curves after the LPM was reported to successfully represent experimental results and predict drying kinetics of fluidized bed drying of different products without vibration (Chen et al., 2018, Ertekin and Firat, 2017, Madhiyanon et al., 2009).

Drying in vibrated fluidized beds has also been successfully modeled with the empirical drying curve approach. Zhao et al. (2014) modeled the drying of lignite in a vibrated fluidized bed under varying process conditions and particle sizes. Perazzini et al. (2017) applied the empirical drying model to the VFB drying of $\gamma$-$Al_2O_3$ particles. They used it to investigate the impact of mechanical vibration on the drying kinetics.

The shape of an empirical drying curve depends on the material properties of the solid, the parameters of the process gas as well as apparatus specifics. Due to the large number of influencing factors, direct comparison of drying kinetics is difficult. Considering different process parameters or different dryer geometries, comparison is almost impossible. Chen et al. (2018) confirmed this by using the empirical drying curve modeling approach for scale-up calculations between different lab-scale fluidized bed dryers, operated with a pharmaceutical excipient of Geldart group A. Their model could successfully predict scale-up at very similar process parameters and plant sizes. However, the model pre-

dictions were less accurate, as the target process parameters deviated stronger from the model development conditions (Chen et al., 2018). Therefore, independent description of material specific evaporation resistances and the influence of process parameters (hydrodynamics and driving force of evaporation) is highly desirable for accurate predictions of drying kinetics, outside process parameters used during the model development.
Stakić and Urošević (2011) considered said material specific resistance in their VFB drying model via the so-called *moisture transport coefficient* or *drying coefficient.* Their material specific, empirical coefficient considered all moisture transport mechanisms inside the solids. It furthermore depended on the material's moisture content and temperature. The influence of vibration on drying kinetics was accounted for by an effective diffusion coefficient. In their model, Stakić and Urošević (2011) used correlations for packed beds to determine mass and heat transfer coefficients. The model represented experimental data of VFB drying of poppy seeds well. However, applicability of this modeling approach for other solids was not investigated (Stakić and Urošević, 2011). Changes in hydrodynamics for other Geldart groups are likely to lower the model's accuracy.
Simple but complete description of material specific drying characteristics is crucial when models shall be compatible with other process condition, dryer geometries or materials. Two approaches have been developed for this purpose: The Characteristic Drying Curve (CDC) and the Reaction Engineering Approach (REA). Both allow for material specific representation of drying kinetics, independent of process parameters. Determination of material specific drying curves only requires a few, well-designed batch drying experiments. Chen and Lin (2004) compared both approaches comprehensively. The CDC and the REA are introduced in detail in the following chapters.

### 2.2.1 Characteristic Drying Curve

The concept of the Characteristic Drying Curve (CDC) was introduced by van Meel (1958). Derived drying curves were later normalized to dimensionless form and applied to fluidized bed drying operations (Burgschweiger et al., 1999). The normalized drying rate $\dot{\nu}$ is plotted against the normalized moisture content $\xi$. This approach is called the Normalized Characteristic Drying Curve (NCDC) (Burgschweiger et al., 1999).
Determination of normalized moisture content requires the equilibrium moisture content $X_{\mathrm{eq}}$ and the critical moisture content $X_{\mathrm{cr}}$. $X_{\mathrm{cr}}$ represents the moisture content at the transition point between first and second drying period:

$$\xi = \frac{X_{\mathrm{i}} - X_{\mathrm{eq}}}{X_{\mathrm{cr}} - X_{\mathrm{eq}}}. \tag{2.24}$$

The NCDC is defined as the ratio of the current evaporation flux and the evaporation flux of the first drying period:

$$\dot{\nu} = \frac{\dot{m}_{\mathrm{v^*,i}}}{\dot{m}_{\mathrm{v^*,1stDP}}}. \tag{2.25}$$

The evaporation flux $\dot{m}_{v^*,i}$ is calculated, using the bed mass of the dry solids $M_{P,dry}$, the change of moisture content over time $dX/dt$ and the evaporation area of the particles $A$:

$$\dot{m}_{v^*,i} = \frac{dX}{dt}\frac{M_{P,dry}}{A} . \tag{2.26}$$

The NCDC method has been successfully used in modeling of fluidized bed drying operations as well as spray drying (Alaathar, 2017, Alaathar et al., 2013, Burgschweiger et al., 1999, Burgschweiger and Tsotsas, 2002, Chen et al., 2017, Langrish and Kockel, 2001).

Material specific drying kinetics after the NCDC, derived by experiments on a single particle, are able to describe fluidized bed drying kinetics of the same material. Likewise, batch drying experiments in fluidized beds may be used for the determination of material specific NCDC (Burgschweiger et al., 1999). A disadvantage of the NCDC is the required knowledge of $X_{cr}$, which depends on the drying conditions. For low driving forces, water at the surface will evaporate at a lower rate and the intraparticle transport phenomena will start to hinder the drying process at a later stage. Thus, $X_{cr}$ is reduced, leading to a different shape of the NCDC (Burgschweiger et al., 1999).

### 2.2.2 Reaction Engineering Approach

The Reaction Engineering Approach (REA) is another approach of representing material specific drying kinetics. Chen and Xie (1997) introduced this approach, based on the principle of chemical reaction kinetics. It has been used successfully in accurate models for single particle drying (Chen and Lin, 2005, Chen and Xie, 1997, Chew et al., 2013, Le et al., 2020) as well as film drying applications (Putranto et al., 2011a,b,c). To the author's best knowledge, this approach has not previously been used in the modeling of fluidized bed drying. The principle of the REA is the representation of the resistance of evaporation by an activation energy. The principle of determination of the REA for fluidized bed drying was previously published by the author (Lehmann et al., 2020) and is described in the following part. Generally, the drying of a solid is described by (Chen and Xie, 1997):

$$\frac{dm}{dt} = M_{P,dry}\frac{dX}{dt} = -\beta \cdot A \cdot (c_{v,s} - c_{v,\infty}) , \tag{2.27}$$

where $dm/dt$ expresses the drying rate. It is equivalent to the product of the mass of the dry solids in the sample $M_{P,dry}$ and the change of moisture content over time $dX/dt$. The drying rate depends on the mass transfer coefficient $\beta$, the surface area of the particles $A$ and the driving force; in this case the difference between the water vapor concentration at the surface of the particle and water vapor concentration in the drying air $(c_{v,s}-c_{v,\infty})$. Determination of the vapor concentration at the surface $c_{v,s}$ is difficult, due to changes during the course of the drying process. Hence, another expression is introduced:

$$c_{v,s} = \Phi \cdot c_{v,sat,s}\left(T_{P,s}\right) , \tag{2.28}$$

where, $c_{\mathrm{V,sat,s}}$ depicts the saturation concentration at the surface of the particle, with surface temperature $T_{\mathrm{P,s}}$. $\Phi$ corresponds to the relative humidity of the air, directly at the surface of the particle.
The temperature of the particle is assumed to be constant in all directions (Chen and Xie, 1997), which is valid for small Biot numbers. Therefore, the particle temperature $T_{\mathrm{P}}$ is used for further calculations. Considering the first drying period, i.e. the particle surface is completely covered with water, $\Phi = 1$, meaning saturated air. In the second drying period, when the particle surface is only partially or not all covered with water, the air at the particle surface is not saturated (i.e. $\Phi < 1$). Using the REA, $\Phi$ is expressed, using an Arrhenius type equation, consisting of the activation energy $\Delta E_{\mathrm{v}}$, representing the overall resistance to evaporation:

$$\Phi = \exp\left(-\frac{\Delta E_{\mathrm{v}}}{R \cdot T_{\mathrm{P}}}\right) . \tag{2.29}$$

Replacing $c_{\mathrm{V,s}}$ in equation 2.27 allows for expression of the activation energy with the following equation:

$$\Delta E_{\mathrm{v}} = -RT_{\mathrm{P}} \cdot \ln\left(\frac{-M_{\mathrm{P,dry}}\frac{dX}{dt}\frac{1}{\beta \cdot A} + c_{\mathrm{V},\infty}}{c_{\mathrm{V,sat,s}}}\right) . \tag{2.30}$$

All parameters in equation 2.30 are obtainable from one, well-performed drying experiment. For comparison of drying processes, the activation energy of evaporation is divided by the activation energy in equilibrium state. The particle temperature in equilibrium equals the gas temperature ($T_{\mathrm{P}} = T_{\infty}$). Thus, the activation energy in equilibrium state is computed with:

$$\Delta E_{\mathrm{v,eq}} = -RT_{\infty} \cdot \ln\left(\frac{c_{\mathrm{V},\infty,\mathrm{eq}}}{c_{\mathrm{V,sat,eq}}}\right) . \tag{2.31}$$

The water vapor concentration of the air depends on the water content $Y$ and the density $\rho_{\mathrm{air,dry}}$ of dry air:

$$c_{\mathrm{V}} = Y \cdot \rho_{\mathrm{air,dry}} \, . \tag{2.32}$$

Assuming ideal gas behavior, the density of the dry air is calculated with the atmospheric pressure $p$ and the gas constants of dry air ($R_{\mathrm{air,dry}} = 287.12\,\mathrm{J\,kg^{-1}\,K^{-1}}$):

$$\rho_{\mathrm{air,dry}} = p\frac{R_{\mathrm{air,dry}}}{T_{\mathrm{air}}} . \tag{2.33}$$

The water content $Y$ is calculated by:

$$Y = \frac{\tilde{M}_{\mathrm{V}}}{\tilde{M}_{\mathrm{air}}} \cdot \frac{\varphi \cdot p_{\mathrm{V,sat}}}{p - \varphi \cdot p_{\mathrm{V,sat}}} , \tag{2.34}$$

with the saturation pressure $p_{\mathrm{V,sat}}$, the relative humidity of the air $\varphi$ and the molar masses $\tilde{M}$ of vapor and air. In the temperature range from 0 to 60 °C, $p_{\mathrm{V,sat}}$ is determined with the following correlation (Tsotsas, 2010):

$$p_{\mathrm{v,sat}}[\mathrm{bar}] = 1 \cdot 10^{-3} \cdot 10^{\left(8.3246 - \frac{1799.73}{\vartheta_{\mathrm{v}}[^{\circ}\mathrm{C}]+238.734}\right)} . \tag{2.35}$$

The mass transfer coefficient $\beta$ is calculated using the Sherwood number $Sh$:

$$Sh = \frac{\beta\, d_{\mathrm{P}}}{D} , \tag{2.36}$$

where $D$ is the diffusion coefficient, which strongly depends on the temperature and is calculated for the binary mixture of water vapor and air (Tsotsas, 2010):

$$D_{\mathrm{H_2O,air}} = \frac{2.252}{p[\mathrm{Pa}]} \cdot \left(\frac{T_{\mathrm{v}}[\mathrm{K}]}{273}\right)^{1.81} . \tag{2.37}$$

The calculation of $Sh$ is conducted with a suitable correlation for the examined case. $Sh$ correlations differ significantly for different processes due to different geometries, e.g. comparing single particle and film drying. For fluidized bed applications, $Sh$ is calculated using the correlations described by Wirth (2010). A basic assumption in this approach is the independence of the mass transfer coefficient from gas velocity in fluidized beds. Furthermore, it is assumed that $Sh$ of a single particle represents $Sh$ of the fluidized bed. $Sh$ of a single particle is defined as (Wirth, 2010):

$$Sh_{\text{single particle}} = 2 + \sqrt{Sh_{\mathrm{lam}}^2 + Sh_{\mathrm{turb}}^2} . \tag{2.38}$$

The laminar and turbulent parts of the Sherwood number are functions of the Schmidt $Sc$ and the Reynolds number $Re$:

$$Sh_{\mathrm{lam}} = 0.664 \cdot \sqrt[3]{Sc}\sqrt{Re_{\mathrm{t}}} , \tag{2.39}$$

$$Sh_{\mathrm{turb}} = \frac{0.037 \cdot Re_{\mathrm{t}}^{0.8} \cdot Sc}{1 + 2.443 \cdot Re_{\mathrm{t}}^{-0.1} \cdot \left(Sc^{2/3} - 1\right)} . \tag{2.40}$$

The Reynolds number of the fluidized bed is approximated with the Reynolds number of a single particle at terminal velocity $Re_{\mathrm{t}}$, which in turn is a function of the Archimedes number $Ar$ (Martin, 2010):

$$Re_{\mathrm{t}} = 18\left(\sqrt{1 + \frac{1}{9}\sqrt{Ar}}\right) - 1 . \tag{2.41}$$

The Schmidt number $Sc$ is:

$$Sc = \frac{\nu}{D} . \tag{2.42}$$

By normalizing the activation energy (equation 2.30) with the activation energy in equilibrium (equation 2.31), the REA delivers the material specific drying kinetic in dependence of the moisture content $(X - X_{\text{eq}})$:

$$\frac{\Delta E_{\text{v}}}{\Delta E_{\text{v,eq}}} = \frac{T_{\text{P}} \cdot \ln\left(\frac{-M_{\text{P,dry}}\frac{dX}{dt} \cdot \frac{1}{\beta \cdot A} + c_{\text{v},\infty}}{c_{\text{v,sat}}}\right)}{T_{\infty} \cdot \ln\left(\frac{c_{\text{v},\infty,\text{eq}}}{c_{\text{v,sat,eq}}}\right)} = f\left(X - X_{\text{eq}}\right) . \tag{2.43}$$

An advantage of the REA over the NCDC is that the critical moisture content is not required, because the REA is able to model the entire drying process with one continuous mathematical function (equation 2.43). Since the REA has not been applied to fluidized bed drying processes prior to this work, its applicability is investigated thoroughly in chapter 4.3.1.

Dimensionless expression of the REA is useful for comparison and evaluation of its validity for fluidized bed drying. Therefore, the material specific function may be derived as a function of the normalized moisture content $\xi$:

$$\frac{\Delta E_{\text{v}}}{\Delta E_{\text{v,eq}}} = f(\xi). \tag{2.44}$$

A downside of this however, is the dependence of the critical moisture content. On the other hand it is advantageous for comparison of different materials as well as results gathered in varying experimental systems. Additionally, REA curves are easily comparable to published NCDC data, in form of the inverted REA:

$$\dot{\nu} = 1 - \frac{\Delta E_{\text{v}}}{\Delta E_{\text{v,eq}}}(\xi) . \tag{2.45}$$

# 3

# Experimental Methods

## 3.1 Powder Characteristics

A wide range of particulate materials is used in this work. An overview of the investigated materials and their respective use in the scope of this thesis is shown in Table 3.1. The applied methods for determination of relevant material properties are introduced in the following chapters. The results are summarized in Table 3.2. The minimum fluidization velocity is covered comprehensively in chapter 3.3.3.

### 3.1.1 Density

For the investigation and characterization of powders, different kinds of densities need to be considered. According to DIN 66137-1:2019-03 (DIN, 2019a), important differentiation are to be made between density of the bulk material and single particles. Furthermore, for different purposes, different definitions of the single particle densities are valid (Stieß, 1995). The relevance and measurement techniques of all densities relevant to this work are introduced in the following sections and shown exemplary in Figure 3.1.

Table 3.1: Overview of investigated materials and respective purpose.

| Material | Purpose of investigation |
|---|---|
| Glass beads | method validation |
| WMP* | hydrodynamics, model validation |
| FCC catalyst | hydrodynamics, drying kinetics, model validation |
| Cellets | hydrodynamics, drying kinetics, model validation |
| $\gamma$-$Al_2O_3$ | hydrodynamics, drying kinetics, model validation |

* whole milk powder, commercially available

### Bulk Density

Any bulk of granular material has a certain porosity due to free spaces between individual particles, as depicted in Figure 3.1(a). The density of a loosely packed bulk material $\rho_b$ is a key parameter in the description of powdered materials as well as for the evaluation of fluidization experiments with regard to bed expansion. For instance, the bulk density of the bed material must be known for estimation of the bed mass by measurement of the fixed bed height.
Additionally, the porosity of the bed, also known as bed voidage, may be determined by setting the bulk density in relation to the single particle density. This value is crucial for the calibration of capacitance probe measurements (see chapter 3.3.1), characterization of bed expansion and fluidization regime.
According to DIN 51705:2001-06 (DIN, 2001), the bulk density is measured by carefully filling a vessel of known volume and measuring the mass of filled powder. The ratio of mass and volume gives the bulk density.

### Apparent Density

The apparent particle density $\rho_{app}$ describes the density of a single particle including all pores, as shown in Figure 3.1(b). This density represents the ratio of mass and volume of the particle as it is present in the flow field of fluidized beds. Hence, $\rho_{app}$ is used for description and estimation of fluidization characteristics.
For the measurement of $\rho_{app}$, a known mass of powder is carefully dispersed in a measurement cylinder with a known volume of Glycerol (>99.5 % p.a. water free). The mass of the powder is divided by the change in volume to derive $\rho_{app}$.

### Skeletal Density

The skeletal density $\rho_{sk}$ describes the volume of the particle structure excluding freely accessible pores (see Figure 3.1 (c)). Pores within the particle structure that cannot be penetrated by gas are included in $\rho_{sk}$. The porosity of the particles can be estimated by comparing the apparent and the skeletal density.
Measurement of $\rho_{sk}$ is conducted with a Helium-Pycnometer *Multi Volume Pycnometer 1305* (*micromeritics*, USA) in accordance to DIN 66137-2:2019-03 (DIN, 2019b). Thereby, a known mass of powder is placed in a chamber with volume $V_{cham}$, flushed with helium gas so that all accessible pore space is filled, and a defined pressure $p_1$ is applied in the sample chamber. Opening a reference volume $V_{ref}$ leads to expansion of the helium and a reduction of pressure $p_2$. The change of pressure correlates to the volume of the sample without the accessible pores. The volume of the sample $V_{Sample}$ can be calculated with the following equation:

$$V_{Sample} = V_{cham} - \frac{V_{ref}}{\frac{p_1}{p_2} - 1}. \tag{3.1}$$

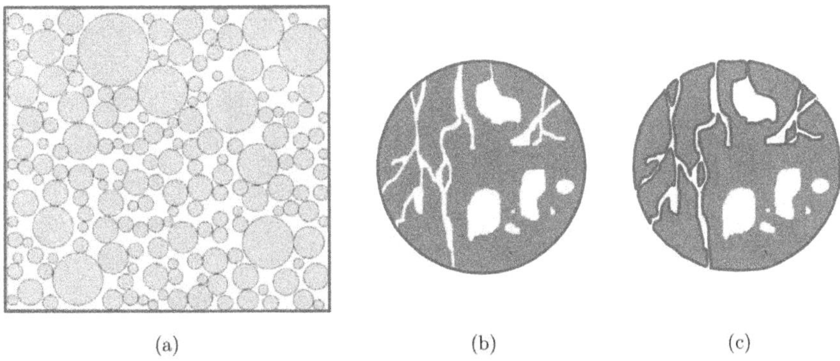


Figure 3.1: Relevant types of particle density in the scope of this work: (a) bulk density, (b) apparent particle density and (c) skeletal particle density; modified from Stieß (1995).

Division of the sample mass by the sample volume results in the skeletal density of the powder.

### 3.1.2 Particle Size and Size Distribution

The particle size distribution (PSD) is determined after an optical measurement approach, using a *CAMSIZER XT* (*Retsch Technology GmbH*, Germany). In this device, the particles pass a stroboscope light source. The shadows of the falling particle are recorded with two cameras (zoom camera and basic camera). The basic camera focuses on particles larger than 100 µm, the zoom camera is for particles smaller than that. The resulting images are analyzed and the particle size distribution is calculated by the device's software. A sketch of the measurement principle is shown in Figure 3.2.
Depending on the properties of the investigated powders, two different configurations of the *CAMSIZER XT* are used. For particles with high flowability (e.g. glass beads), the free fall method is used. Herein, the particles are conveyed on a vibrating chute and fall past the cameras, solely due to gravity.
For more cohesive particles, such as the investigated WMP, the cohesiveness of the particles causes clustering of particles or temporary agglomeration. Hence, the particles are accelerated by pressurized air before passing the cameras. This dispersion method results in breakage of unwanted agglomerates or clusters of particles. Said temporary agglomerates may form during storage or transport of the powder.
The investigated particles are plotted in the Molerus diagram to identify their respective Geldart group, based on particle size and density, in Figure 3.6.
Simultaneously to the particle size, the particle shape is analyzed by the *CAMSIZER XT* in form of the sphericity $\Psi$. According to DIN ISO 9276-6:2012-01 (DIN, 2012), the

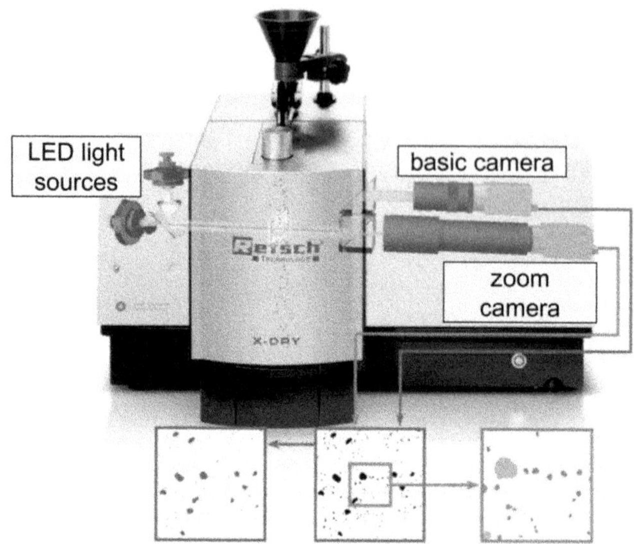


Figure 3.2: Measurement principle of *CAMSIZER XT* (Retsch Technology GmbH, 2017).

sphericity is defined according to Wadell (1933):

$$\Psi_{\text{Wadell}} = \left(\frac{d_{\text{vol}}}{d_{\text{surf}}}\right)^2, \tag{3.2}$$

where $d_{\text{vol}}$ is the volume-equivalent diameter and $d_{\text{surf}}$ is the surface-area-equivalent diameter of the investigated particle. A value of $\Psi_{\text{Wadell}} = 1$ represents an ideal sphere. Lower values describe shapes deviating from an ideal sphere.

### 3.1.3 Moisture Content

The moisture content of the investigated materials during the fluidized bed experiment is determined using a thermo-gravimetric moisture analyzer *EM 120-HR* (*Precisa*, Switzerland). A small sample of the material (approximately 1 g) is heated via infrared radiation to 105 °C, while the mass of the sample is monitored constantly over time. Evaporation of moisture is considered to be the only cause for changes in mass. Hence, the moisture content of the sample is determined. If not stated otherwise, the term moisture content refers to the water content of particles in the scope of this work. Measurement with the *EM 120-HR* takes between seven to ten minutes. During the drying experiments of this work, samples are taken every five to ten minutes. This way, storage of samples under conditions, differing from the drying conditions, is limited to a minimum and reduction of bed mass is minimized.

During some drying experiments, inline measurement of particle moisture content is conducted with an inline capacitance moisture analyzer *SampIn* (*AVA*, Germany). It uses the difference in relative dielectric constant of water and investigated solid materials. The sensor head, containing the sample, consists of the two electrodes of a capacitor. After careful calibration for the respective material in the relevant range of moisture content, the moisture content of the particles in the measurement volume is determined by measurement of the voltage signals of the capacitor electrodes.

The *SampIn* moisture analyzer allows for measurements of particle moisture content with higher time resolution. One measurement is conducted approximately every ten seconds and all measurements over a period of two minutes are averaged, resulting in one data point, used for analysis. The *SampIn* device is only applicable in drying experiments in the lab-scale fluidized bed dryer, because the inline measurement of particle moisture content requires the material to be stationary for a few seconds in the sensor head. After each measurement, the sensor head is emptied by a short pressure impulse of dried and compressed air.

### 3.1.4 Sorption Isotherms

Knowledge of the particles' moisture content in equilibrium $X_{\mathrm{eq}}$ with the surrounding gas is of utmost importance for the prediction and simulation of the drying process. $X_{\mathrm{eq}}$ is reached when adsorption and desorption processes of moisture between gas and particles are in equilibrium. The equilibrium moisture content depends on the material structure, such as porosity, pore size, pore structure and turtuosity, as well as the temperature and relative humidity of the surrounding gas (Tsotsas et al., 2000).

In order to describe the adsorption behavior of a powdered material, the equilibrium moisture content is measured at constant temperature and a range of relative humidity $\varphi$. Therein, the mass of the sample is monitored and $\varphi$ is kept constant until equilibrium is reached, i.e. the sample mass does not change due to adsorption of moisture. After the equilibrium is reached, $\varphi$ is increased until equilibrium is reached again. This step-wise procedure is repeated for several levels of $\varphi$ and the resulting equilibrium moisture contents are plotted against $\varphi$, resulting in an adsorption isotherm curve of the investigated material (Surface Measurement Systems Ltd., 2015, Tsotsas et al., 2000).

Applying the same procedure, but starting with high relative humidity and decreasing it step-wise, results in the desorption isotherm. The desorption isotherm characterizes the theoretical limit of the drying process under the given conditions. Desorption isotherms of the materials used in the drying experiments are therefore measured for a range of temperatures.

The shape of the sorption isotherms is related to the underlying adsorption and desorption processes of water molecules inside the particles. Specifically, the pore size plays a significant role. The International Union of Pure and Applied Chemistry (IUPAC) defined six types of sorption isotherms. Detailed discussions of the underlying phenomena are conducted in AlOthman (2012) and Tsotsas et al. (2000). To describe the physical processes relevant to this thesis, Figure 3.3 shows an exemplary adsorption isotherm and

the underlying adsorption mechanisms. At low relative humidity, water molecules are adsorbed at the surface of the pores in a single layer (A). Molecular forces, such as van der-Waals forces, dipole or hydrogen bonds, dominate the adsorption. The equilibrium moisture content increases with increasing relative humidity. The more developed the layer of water molecules on the particle surface, the slower the increase of equilibrium moisture content. This adsorption regime is also called mono-layer or Langmuir adsorption (Langmuir, 1932).
Further increase in relative humidity results in adsorption of additional molecules in a second, or more, layers. This is called multi-layer adsorption (see Figure 3.3 B) (Brunauer et al., 1938). At high relative humidity, the layers of water molecules at the walls of the pores may grow to such an extent, that they start influencing each other. Hereby, the adsorbed molecules form concave menisci inside the pores. The vapor pressure $p_v$' at the concave menisci is lower than the surrounding vapor pressure $p_v$, resulting in capillary condensation (see Figure 3.3 C). This enhances the adsorption of further molecules and leads to a strong increase in equilibrium moisture content (Tsotsas et al., 2000).
Measurement of the sorption behavior is conducted in a *DVS Resolution* (*Surface Measurement Systems Ldt.*, UK), using the Dynamic Vapor Sorption (DVS) principle. Hereby, the particle sample is positioned in a flow of conditioned gas. Dried nitrogen is used in the scope of this work. Moisture content and temperature are controlled, while the mass of the sample is monitored via microbalance. Figure 3.4 shows a photograph of the *DVS Resolution* and a simplified flow chart of the device. The dynamic method reduces the measurement time of isotherms by several days or weeks, compared to static methods, where the sample is placed in stationary conditioned gas.
Adsorption and desorption isotherms are measured for the investigated particles in the relevant range of drying temperatures. Desorption isotherms are crucial in the design and modeling of drying processes. Hence, only the measured desorption isotherms are shown in Figure 3.5 and used in the simulations in chapter 5. Repeated measurement of ad- and desorption cycles with the same sample confirm reversibility, i.e. neither changes to the materials structure nor composition are evident for FCC catalyst, Cellets and $\gamma$-$Al_2O_3$ particles.

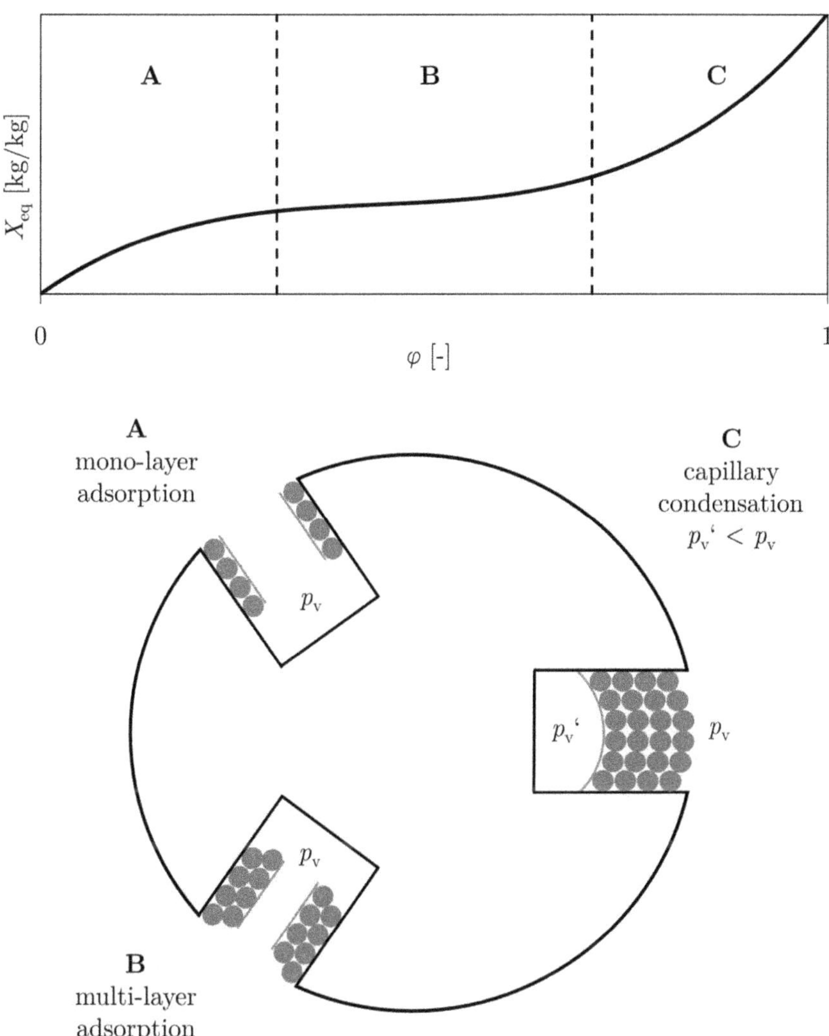


Figure 3.3: Exemplary adsorption isotherm (top) and schematic of the most common underlying adsorption mechanisms in particle pores (bottom); A: mono-layer or Langmuir adsorption, B: multi-layer adsorption, C: capillary condensation; including vapor pressure $p_v$ and vapor pressure at concave menisci of capillaries $p_v$'.

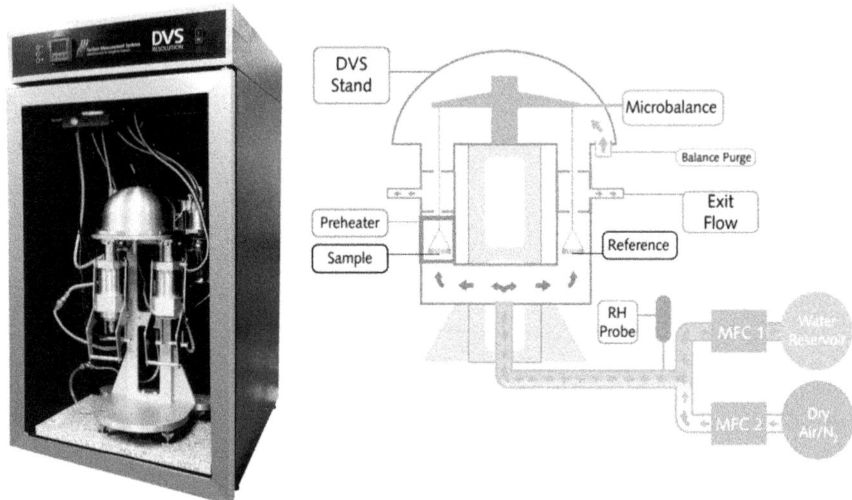


Figure 3.4: Photograph of the *DVS Resolution* (left) and schematic flow chart of the measurement principle (right) (Surface Measurement Systems Ltd., 2015).

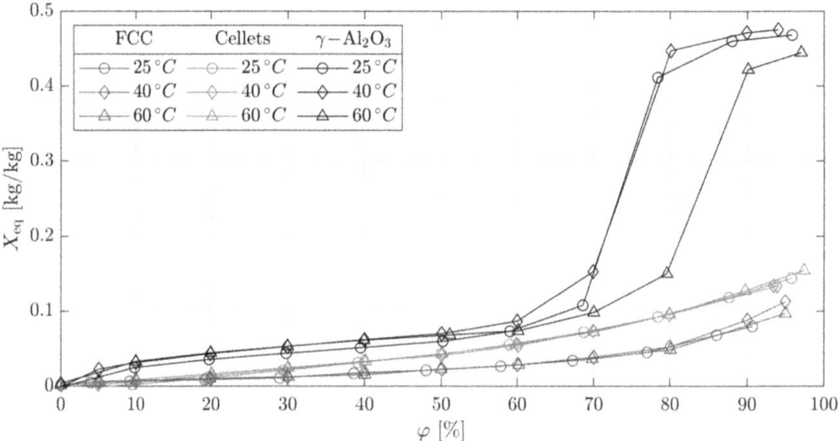


Figure 3.5: Desorption isotherms of FCC catalyst, Cellets and $\gamma$-$Al_2O_3$ particles at 25 °C, 40 °C and 60 °C.

### 3.1.5 Specific Surface Area

Measurement of the specific surface area of solids is conducted by using the BET principle. It is named after it's inventors, Brunauer, Emment and Teller. It uses the adsorption of vapor on the solids' surfaces in the mono-layer regime (see Figure 3.3). Vapor pressure $p_\mathrm{v}$ of the adsorbed vapor in the carrier gas and volume of adsorbed vapor $V$ are monitored during the measurement (Brunauer et al., 1938). Utilization of non-polar, organic fluids, that do not dissolve the investigated solids, allows for measurement of the specific surface area at ambient temperature in the *DVS Resolution* (Surface Measurement Systems Ltd., 2015).

In the scope of this work, n-Octane is used as the adsorbed vapor phase and dried nitrogen is used as carrier gas. Adsorption isotherms at 25 °C in the mono-layer adsorption regime (relative humidity from 5 to 30 %) are determined. Applying the BET equation:

$$\frac{1}{V}\frac{p_{\mathrm{v}}}{1-p_{\mathrm{v}}} = \frac{c-1}{cV_{\mathrm{m}}}p_{\mathrm{v}} + \frac{1}{cV_{\mathrm{m}}}, \tag{3.3}$$

and plotting the left hand side against the relative humidity ($\varphi = p_{\mathrm{v}}/p_0$), results in a straight line. Based on the slope and the intercept, the constants $c$ and $V_{\mathrm{m}}$, of which the latter is the volume of adsorbed gas in mono-layers, are calculated. The specific surface area $S$ is calculated with knowledge of the molar mass of the adsorbed vapor phase $\tilde{M}$, the average cross sectional area of the adsorbed molecules $A_{\mathrm{m}}$ and the Avogadro constant $N_{\mathrm{A}}$ (Brunauer et al., 1938, Surface Measurement Systems Ltd., 2015):

$$S = \frac{V_{\mathrm{m}} \cdot N_{\mathrm{A}} \cdot A_{\mathrm{m}}}{\tilde{M}}. \tag{3.4}$$

Table 3.2: Overview of relevant properties of powder materials used in experiments and simulations.

| Material | Skeletal density [kg m$^{-3}$] | Apparent density [kg m$^{-3}$] | Bulk density [kg m$^{-3}$] | Sauter diameter [$\mu m$] | Sphericity [-] | Specific surface area [m$^2$ g$^{-1}$] | Geldart group [-] | $u_{mf}$* [m s$^{-1}$] |
|---|---|---|---|---|---|---|---|---|
| Glass beads | 2500.0 | 2500.0 | 1514.1 | 63.3 | 0.91 | <0.01 | A | 0.008 |
| WMP | 1256.0 | 866.0 | 404.0 | 117.5 | 0.72 | 3.7 | A/C | 0.1 |
| FCC catalyst | 2729.2 | 1632.5 | 774.0 | 41.6 | 0.86 | 93.9 | A | 0.015 |
| Cellets 500 | 1464.6 | 1113.6 | 826.1 | 631.8 | 0.85 | 63.6 | B | 0.2 |
| $\gamma$-$Al_2O_3$ | 3261.2 | 1006.4 | 590.2 | 1701.4 | 0.97 | 206.3 | D | 0.51 |

* measured with dry powder and non-vibrated fluidization, using dried pressurized air

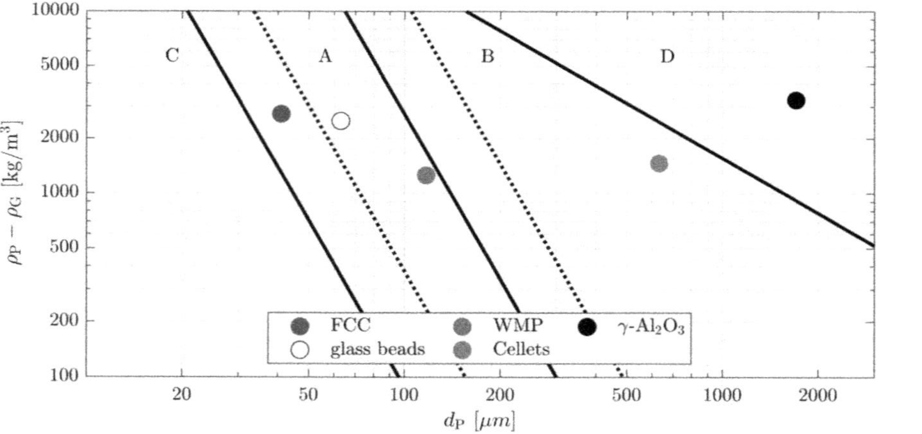


Figure 3.6: Molerus diagram, showing the Geldart groups of the investigated powder materials.

## 3.2 Preparation of Experiments

For investigation of hydrodynamics in dependence of moisture content as well as drying in fluidized beds, the moisture content of the particles needs to be adjusted. Particles, insoluble in water (FCC catalyst, Cellets and $\gamma$-$Al_2O_3$), are mixed with demineralized water by hand, before each experiment, to set the desired moisture content. Visual observation of the mixing suggested uniform mixing after approximately five minutes. In order to ensure uniform mixing and to achieve a narrow moisture distribution, particles are mixed for at least ten minutes.
In case of water-soluble whole milk powder (WMP), humidification of the powder requires different measures. The WMP is fluidized with moist air in the pilot-plant-scale fluidized bed (more details in chapter 3.3). During this, WMP adsorbs moisture from the fluidization air. However, the composition of the WMP poses additional challenges. The stickiness of the WMP increases with increasing moisture content. Additionally, glass transition of the WMP is to be avoided, because it is accompanied by strong increase in stickiness, leading to complete, irreversible and almost instant defluidization of the bed (Murti et al., 2010, Palzer, 2010, Vuataz, 2002). Consequently, the upper limit of moisture contents of WMP is $0.09\,\mathrm{kg\,kg^{-1}}$.

## 3.3 Pilot-Plant-Scale Vibrated Fluidized Bed Dryer

A pilot-plant-scale vibrated fluidized bed (VFB) dryer is designed and custom-built for the experimental investigations in this work. It has a rectangular cross section of 250 mm by 500 mm and the fluidization chamber measures two meters in height. Key features of the dryer are shown in Figure 3.7. Visual observation of the bed can be made, since the front wall of the dryer is made of glass. The bed height during operation is measured, using a ruler that is mounted to the glass wall. The corresponding bed porosity $\varepsilon$ is calculated, using the known bed mass $M_\mathrm{P}$, cross sectional area of the dryer $A$ and apparent particle density $\rho_\mathrm{app}$:

$$\varepsilon = 1 - \frac{M_\mathrm{P}}{A \cdot H_\mathrm{fb} \cdot \rho_{app}} \,. \tag{3.5}$$

The vibrating parts of the dryer (windbox, distributor plate and drying chamber) are suspended on springs. The expansion zone is decoupled from the vibrating parts via rubber gaiter. The distributor plate consists of a multilayered, stainless steel, wire mesh (*Spörl oHG*, Germany) with a porosity of 15 % and a perforated plate with large hole diameter (17 mm) for mechanical support.
The dryer can be operated with and without vibration. The range of vibration parameters (frequency $f$ and amplitude $A_\mathrm{vib}$) is designed similar to typical parameters in industrial WMP drying (Pisecký, 2012). The investigated parameter combinations are summarized in Table 3.3. Vibration is introduced by two synchronized eccentric motors

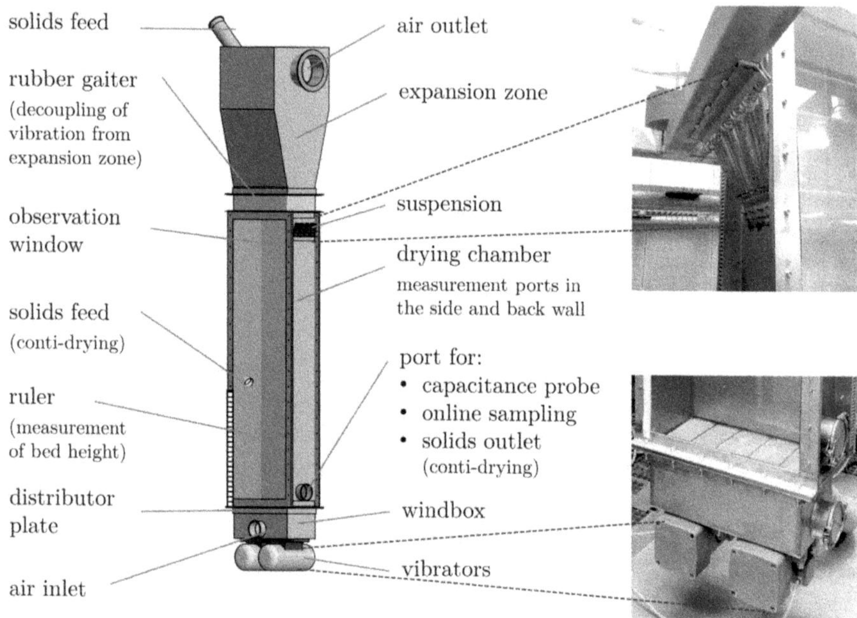


Figure 3.7: Features of the custom-built vibrated fluidized bed (VFB) dryer; drying chamber measuring 250 by 500 mm and two meters in height. Strictly vertical vibration is introduced by synchronized eccentric motors; frequency is controlled via frequency converter, amplitude results from frequency and strength and number of springs in suspension of the VFB. Ports for pressure and temperature measurements are located in the back wall of the drying chamber.

(*Würges*, Germany), mounted to the bottom of the windbox. Strictly vertical vibration is introduced to the bed due to the orientation of the eccentric motors. The frequency of vibration is controlled via frequency inverter. The amplitude is a result of the applied frequency and the number and strength of springs in the suspension of the VFB dryer. Both, frequency and amplitude of vibration are measured with a pen recorder.

Figure 3.8 shows a simplified flow chart of the VFB dryer. Air is used as fluidization gas. Temperature and moisture content of the drying air are adjusted in an air conditioning unit. It consists of two electric heaters and a humidifier. All three parts are controlled individually. The drying air passes the pre-heater first, then the humidifier and lastly, the second heater. In the humidifier, demineralized water is sprayed into the air stream. The air also passes over liquid water in the humidifier's pit, thereby almost reaching saturation. The final temperature and relative humidity are set with the second heater. Dried, pressurized air is used as fluidization air for superficial gas velocities up to $0.3\,\mathrm{m\,s^{-1}}$. In this case, the volume flow rate is set with mass flow controllers (*Bronkhorst*, Germany). For higher gas flow rates (up to $0.7\,\mathrm{m\,s^{-1}}$), a rotary blower is used, the dryer

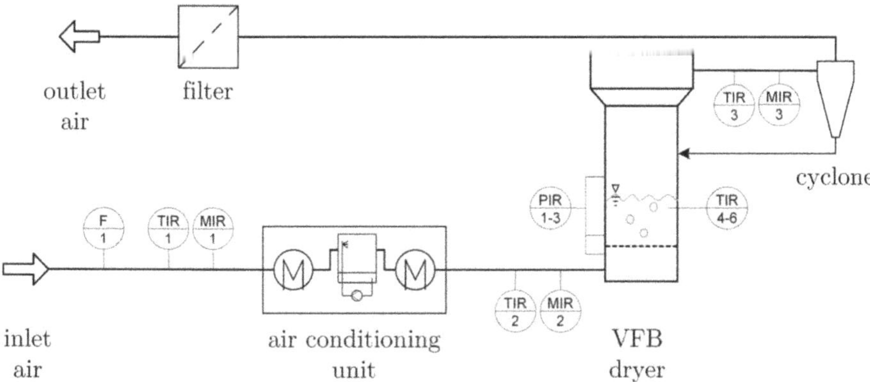


Figure 3.8: Simplified flowchart of the VFB dryer; including air conditioning unit (consisting of a pre-heater, humidifier and second heater). Furthermore, showing key measurement locations of temperature (TIR), humidity (MIR) and flow rate (F) of drying air, bed temperature (TIR 4-6) and pressure drop across bed and distributor (PIR 1-3), as well as recirculation of entrained particles via cyclone and exhaust gas treatment.

Table 3.3: Overview of investigated vibration parameters.

| | | | | | | |
|---|---|---|---|---|---|---|
| frequency $f$ | [Hz] | 0 | 4 | 6 | 8 | 10 |
| amplitude $A_{vib}$ | [mm] | 0 | 5 | 4 | 3.5 | 3 |
| vibration intensity $\Lambda$ | [-] | 0 | 0.32 | 0.59 | 0.9 | 1.21 |

is operated with ambient air and the flow rate is measured via an orifice plate, according to DIN EN ISO 5167-2:2003 (DIN, 2003). In either case, relative humidity and temperature of the process air are measured before the air conditioning unit, at the entrance of the windbox and at the exit of the drying chamber. Particle and gas temperatures are measured with PT100 thermometers at several positions inside and above the bed. Data acquisition is conducted in LabView.
The exhaust air from the VFB dryer is led through a cyclone in order to recycle entrained particles into the dryer or collect them externally. After passing through a filter, the air is released to the environment. A detailed flowchart of the VFB dryer's periphery and measurement system is shown in Figure A.1. The installed measurement equipment for investigation of fluidized bed hydrodynamics and bubble characteristics is introduced in the following chapter.

### 3.3.1 Measurement Equipment

As introduced in chapter 2.1.2, different techniques have been used successfully to investigate the hydrodynamics of fluidized beds. Pressure measurements are conducted in the scope of this work. Due to the size of the VFB, investigation of bubble characteristics is only feasible via local intrusive measurements (via capacitance probe) and global image analysis (via high speed camera) through the glass wall of the dryer.

### Pressure Drop Measurement

Pressure readings are taken at different positions in the dryer with relative pressure sensors HXCM (*Sensor Technics*, Germany). Relative pressure drop is measured across the distributor and five height intervals in the fluidized bed as well as the freeboard. A detailed list of all installed sensors is provided in Appendix A. The pressure signal is used to determine the onset of fluidization ($u_{\mathrm{mf}}$ and $u_{\mathrm{cf}}$), as introduced in chapter 2.1.

### Capacitance Probe

An in-house built capacitance probe is used in this work to investigate local bubble characteristics inside the VFB. It uses the dependency of the relative dielectric constant from the material between two electrodes of a capacitor. If, for instance, particles pass the field lines of a capacitor the dielectric constant $K_{\mathrm{i}}$, and thereby the capacitance, is affected. The capacitance is measured via voltage signals. Thus, the local solid concentration $c_{\mathrm{P}}$ is determined inside a fluidized bed. The solids concentration $c_{\mathrm{P,fix}}$ as well as the relative dielectric constants of the fixed bed $K_{\mathrm{fix}}$ and solid free state $K_{\mathrm{pf}}$ must be known in order to calibrate the sensor. The local solid concentration $c_{\mathrm{P}}$ is then calculated with the following equation (Wiesendorf, 2000, Wiesendorf and Werther, 2000):

$$c_{\mathrm{P}} = c_{\mathrm{P,fix}} \cdot \frac{K_{\mathrm{e}} - K_{\mathrm{pf}}}{K_{\mathrm{fix}} - K_{\mathrm{pf}} \cdot \left(1 + \frac{(K_{\mathrm{fix}} - K_{\mathrm{f}}) \cdot (K_{\mathrm{fix}} - 3)}{K_{\mathrm{fix}} \cdot K_{\mathrm{pf}}}\right)} \tag{3.6}$$

Application of two of the described capacitors (also referred to as channels) in vertical alignment with known distance allows for determination of velocities of passing bubbles (in bubbling fluidized beds) or of passing particles in more dilute systems (e.g. the freeboard). The time delay between the signals of the two channels is analyzed using cross correlation. The velocity of passing bubbles is then deduced from the distance between the two needles and the time delay of the signals (Wiesendorf and Werther, 2000, Wytrwat et al., 2020). Figure 3.9 shows the setup of the two-channel capacitance probe, used in this work. It was also used by Wytrwat et al. (2020). Each capacitor/channel consists of a cylindrical core electrode (needle) and a ring electrode around it. The capacitance probe is inserted into the VFB through a socket in the side wall. The horizontal position of the probe can be adjusted continuously along the entire width and depth of the drying chamber.

The measurements are conducted at a sampling frequency of 10 000 Hz. Data is recorded for 60 s at each investigated position and gas velocity, respectively. Calibration of the capacitance probe is done in-situ. Hereby, the highest measured voltage over the entire measurement period is assumed to represent the fixed bed. The lowest measured voltage on the other hand, corresponds to the solid free state ($c_{\mathrm{P}} = 0$). This calibration method and the fact that bubbles are never totally free of solids, especially in the borders to the

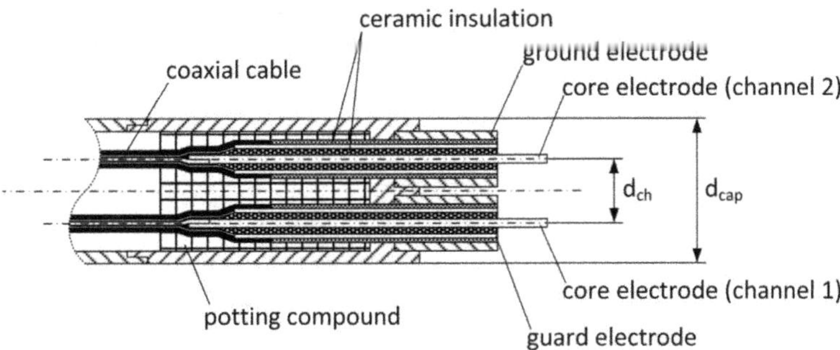


Figure 3.9: Design of in-house built capacitance probe; measuring $d_{\text{cap}} = 16\,\text{mm}$ in diameter and distance between the two core electrodes $d_{\text{ch}} = 5.9\,\text{mm}$; adopted from Wytrwat et al. (2020).

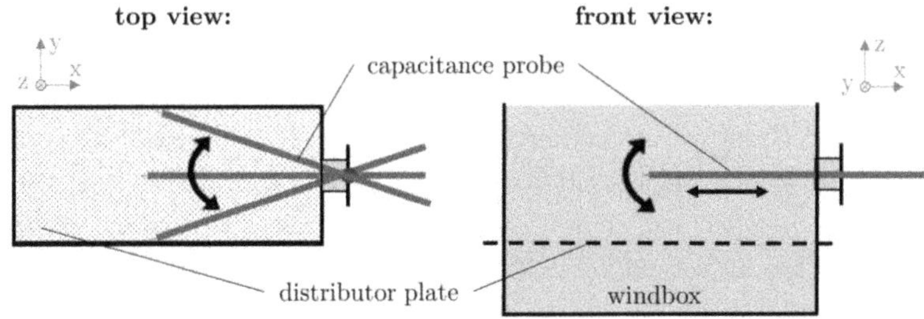


Figure 3.10: Adjustable position of the capacitance probe inside the VFB dyer; top view (left) and front view, through the glass wall (right).

suspension phase, require determination of a threshold value. This threshold defines the solids concentration $c_{\text{P,bub}}$ at the transition from bubble to suspension phase. The value of $c_{\text{P,bub}}$ is determined by validating the capacitance signals, measured at the glass wall, with data from high speed images. The bubble volume fraction is the ratio of all data points with $c_{\text{P}} < c_{\text{P,bub}}$ and the total number of data points.

The capacitance probe is inserted into the VFB through the sockets in the narrow walls. Lateral and vertical position of the probe can be adjusted, as shown schematically in Figure 3.10. Capacitance measurements are conducted in the center of the bed and close to the glass wall (without touching the wall).

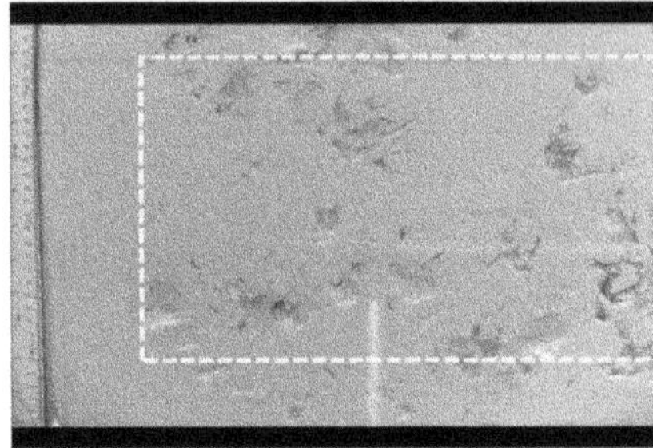 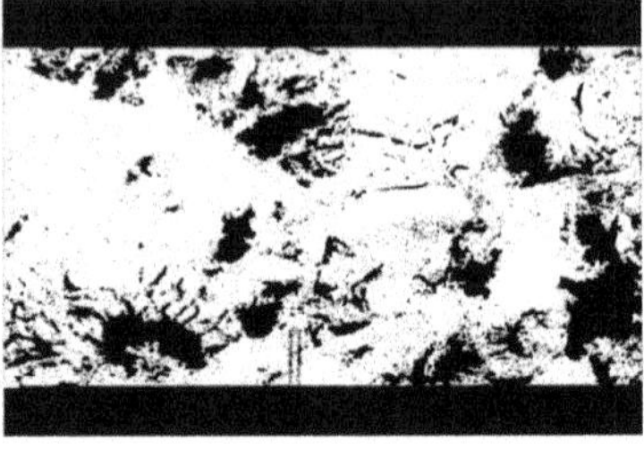

Figure 3.11: Comparison of original gray scale image (left) and segmented binary image (right); showing suspension phase (white pixels) and solid free bubbles and cracks (black pixels). Image previously published in Lehmann et al. (2019).

#### High Speed Camera

The transparent front wall of the VFB dryer allows for visual observation of the bed. Investigations of bubble characteristics via high speed camera (IDT-NX 4-S2 *Imaging Solutions GmbH*, Germany) are conducted. Gray scale images are recorded at 100 fps for a minimum of 30 s. After segmentation of the gray scale images in MATLAB the bubble volume fraction is calculated. Hereby, black pixels represent the bubble phase, white pixels designate the suspension phase, respectively. An example is shown in Figure 3.11. As a simplifying approach, it is assumed that the glass wall acts as a cut through the isometric fluidized bed. The validity of this approach is established in chapter 3.3.5.

### 3.3.2 Modes of Operation

Investigations in the VFB dryer are performed in batch and continuous operation. Batch operations are conducted for investigations of hydrodynamics as well as drying kinetics and determination of material specific drying curves of the investigated materials. Continuous operation is used for investigation of residence time characteristics. Furthermore, continuous drying is conducted to gather extensive data for validation of the flowsheet simulation model. Details of the respective settings and configurations of the VFB dryer are introduced in the following sections.

#### Batch Operation

Hydrodynamic characteristics are investigated in batch operation of the VFB. A schematic flowchart of the VFB in batch operation is depicted in Figure 3.12. Entrained particles are recirculated into the dryer via cyclone in order to keep the bed mass as constant as possible during the investigations. The lower limit of fluidization is investigated in batch mode. Here, the influence of vibration and particle moisture content on the minimum fluidization velocity is of special interest and discussed in chapter 3.3.3. Additionally, bubble characteristics and overall bed expansion are investigated, using high speed cam-

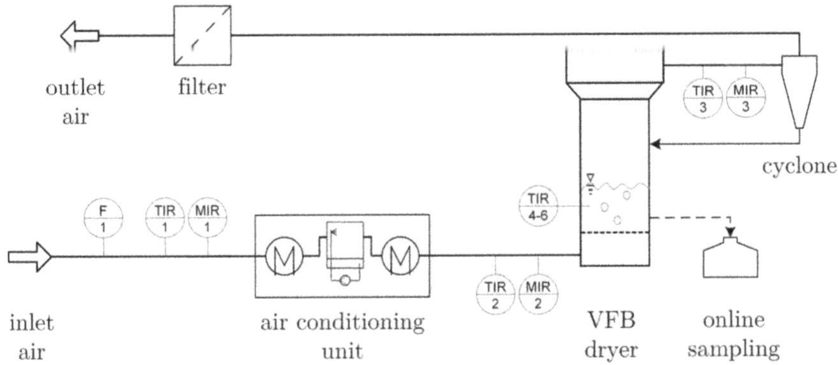


Figure 3.12: Simplified flowchart of VFB dryer in batch operation; showing key measurement locations of temperature (TIR), humidity (MIR) and flow rate (F) of drying air and bed temperature (TIR 4-6). Samples are taken from inside the drying chamber at a height of 10 cm above the distributor to monitor particle moisture content.

era and capacitance probe (see chapter 3.3.5). During batch operations, samples can be taken from the bed via a retractable sampling probe. A schematic of the sampling probe is shown in Figure 3.13. Samples are analyzed externally regarding their moisture content and particle size distribution.

The sampling probe is of particular importance during batch drying experiments. With it, particle moisture content is monitored over time and the drying kinetics are investigated. Details about the conducted parameter studies, analysis and interpretation of the results are covered in chapter 4.3.

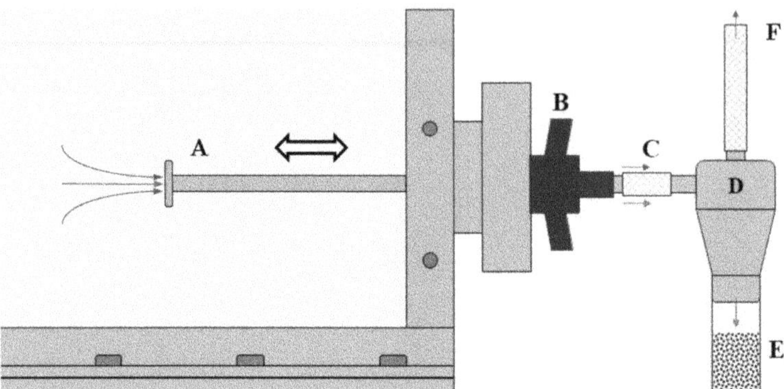


Figure 3.13: Schematic of the retractable sampling probe; consisting of A: steel pipe, B: connection piece, C: connection hose, D: cyclone, E: sample container, F: connection hose to vacuum source.

### Continuous Operation

During continuous operation, feed particles are stored in a bunker and fed into the drying chamber via rotary valve. The mass flow of the particles is controlled by adjusting the speed of the rotary valve. The feed enters the drying chamber through the narrow side of the VFB, approximately 40 cm above the distributor. Particles leave the dryer at the opposite side through an outflow hole. The height of the outflow hole is varied between 8 cm and 16 cm, thus allowing for investigation of the influence of bed height on particle residence time and overall drying process. A schematic flowchart of the VFB dryer under continuous operation is shown in Figure 3.14. Continuous drying experiments are conducted for a variety of process parameters and materials, for validation the flowsheet simulation model. These experiments are discussed in detail in chapter 5. Besides drying experiments, the residence time distribution of the particles is investigated, using the continuous setting. Details about the experimental investigations are introduced in chapter 3.3.6. The results are discussed in chapter 4.2.

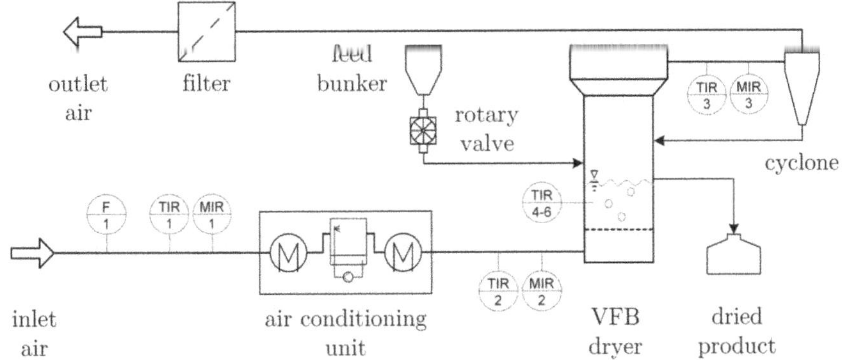


Figure 3.14: Simplified flowchart of VFB dryer in continuous drying operation; showing key measurement locations of temperature (TIR), humidity (MIR) and flow rate (F) and bed temperature (TIR 4-6). Feed is stored in a bunker and conveyed into the dryer via rotary valve, also controlling the feed mass flow. Dried particles exit the dryer at the opposite side through an outflow hole with variable height between 8 and 16 cm.

### 3.3.3 Minimum Fluidization Velocity

The basic principle of minimum fluidization velocity $u_{\mathrm{mf}}$ is introduced in chapter 2.1. Experimental determination is performed via pressure drop measurements across the bed and plotted against gas velocity. The intercept of linearly extrapolations of measured pressure drop curves in the fixed bed and the fluidized state is defined as $u_{\mathrm{mf}}$ (see Figure 2.1).

The $u_{\mathrm{mf}}$ value is an important parameter in the modeling of fluidized bed hydrodynamics as well as drying kinetics (Kunii et al., 2013, Tsotsas et al., 2000). Consequently, detailed knowledge of $u_{\mathrm{mf}}$ of the investigated powders at the respective operation parameters (vibration, moisture content) is crucial and therefore investigated.

Recently, McLaren et al. (2021) proposed a correlation for prediction of $u_{\mathrm{mf}}$ of vibrated fluidized beds of Geldart B and D particles. Their empirical correlation for the Reynolds number at minimum fluidization $Re_{\mathrm{mf}}$ was derived via dimension analysis. It showed accurate predictions for their own measurements as well as literature data (McLaren et al., 2021).

$$Re_{\mathrm{mf}} = 0.075 Ar^{0.91} \cdot \varepsilon_{\mathrm{mf}}^{2.373} \cdot \Lambda^{0.074} \cdot \left(\frac{2\pi f \Delta\rho\, d_{\mathrm{P}}^2}{\eta_{\mathrm{G}}}\right)^{-0.257} \cdot \left(\frac{A_{\mathrm{vib}}}{d_{\mathrm{P}}}\right)^{-0.123} \tag{3.7}$$

This correlation was developed for fluidization of dry particles, i.e. the presence of cohesive forces in form of liquid bridges is not considered. Furthermore, electrostatic effects were minimized during the measurements (McLaren et al., 2021). To the author's best knowledge, this is the only available correlation for $u_{\mathrm{mf}}$ of vibrated fluidized beds. Predictions from McLaren's model are compared to measurements of this work within this chapter.

Determination of $u_{mf}$ is conducted based on pressure drop measurements. The pressure drop across the distributor plate, measured with an empty bed, is subtracted from the overall pressure drop (of bed plus distributor) and then plotted against gas velocity. Pressure drop measurements are recorded at 100 Hz for 30 s. Average values of measured pressure drop and gas velocity are calculated over each time interval and treated as individual data points in the $u_{mf}$-plots. Pressure drop measurements are conducted in the fully fluidized regime without causing entrainment of particles. The gas velocity is decreased step-wise through the fluidized state and the fixed bed regime.

Special focus lies on the influence of vibration and particle moisture content on $u_{mf}$. Thus, particle moisture content is set to a desired value, and $u_{mf}$ measurements are conducted for different vibration parameters (see Table 3.3). Subsequently, the particles are dried gently until the next target moisture content is reached and the procedure is repeated. Investigation of each material consists of four different moisture levels and five sets of vibration parameters each. Small samples (<20 g in total) are taken in ten-minute intervals over the course of the investigation. The samples' moisture content is measured. In all cases, the moisture content is confirmed constant (within $\pm 0.005$ kg/kg the target moisture content).

The results for FCC catalyst are shown in Figure 3.15. The measured $u_{mf}$ of FCC is plotted against vibration intensity $\Lambda$ for four levels of particle moisture content. Evidently, raising the moisture content results in larger $u_{mf}$. Increasing moisture content leads to higher cohesive forces, which need to be overcome by the gas to achieve fluidization. The introduction of vibration leads to a decrease in $u_{mf}$ at a vibration frequency of 4 Hz ($\Lambda = 0.32$), for the highest investigated moisture content ($X = 0.3\,\mathrm{kg\,kg^{-1}}$). With the next higher vibration intensity (vibration frequency 6 Hz), $u_{mf}$ increases again to the level of the non-vibrated case. Decreasing $u_{mf}$ is observed for any further increase in vibration intensity. The same trend is evident for all investigated moisture content levels. It may be explained by stronger reduction of cohesive forces and thus $u_{mf}$, at higher vibration intensities. The resonance frequency of the investigated particle bed is computed to approximately 4 Hz with equation 2.2. Barletta et al. (2013) reported a particularly strong impact of vibration at the resonance frequency, i.e. stronger reduction of $u_{mf}$, compared to other vibration frequencies. This observation coincides with the measurements presented here.

Furthermore, Figure 3.15 shows that the reduction of $u_{mf}$ is more pronounced when the moisture content and respectively the cohesive forces are higher. A reduction of approximately 25 %, 22 % and 13 % of $u_{mf}$ for $0.3\,\mathrm{kg\,kg^{-1}}$, $0.18\,\mathrm{kg\,kg^{-1}}$ and $0.1\,\mathrm{kg\,kg^{-1}}$ moisture content, respectively is caused by a vibration intensity of 1.21, compared to the non-vibrated case. Interestingly, for the highest vibration intensity, the values of $u_{mf}$ are very similar, although the moisture content differs significantly. This leads to the conclusion that cohesive forces, introduced by moisture, are almost completely overcome by the vibration and that the powder is fluidized like the dry powder.

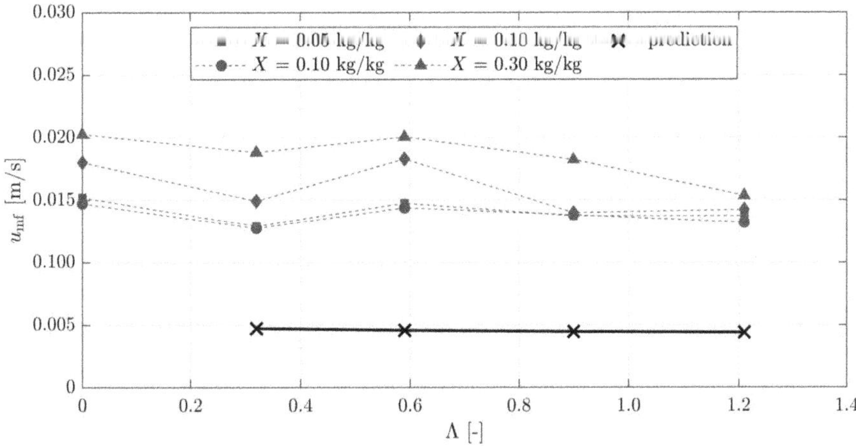


Figure 3.15: Measured minimum fluidization velocity ($u_{mf}$) of FCC catalyst against vibration intensity ($\Lambda$), showing influence of particle moisture content ($X$) and predictions after McLaren et al. (2021).

Besides the measured $u_{mf}$, Figure 3.15 also shows predictions according to McLaren's correlation (equation 3.7). The correlation under-predicts the measurements significantly. Since it was developed for Geldart B and D particles and as the investigated FCC catalyst is of Geldart group A, deviations between predictions and measurements are not surprising. McLaren's correlation does not include cohesive effects of any kind. For Geldart A particles however, cohesive effects play a significant role in their fluidization. Consequently, the model development and simulation of vibrated fluidized bed dryers of this work (see chapters 4 and 5 respectively) rely on the measured values.

The $u_{mf}$ measurements of Cellets are plotted in Figure 3.16, showing the same trend regarding the strongest reduction of $u_{mf}$ at the resonance frequency of the bed around 4 Hz. Reduction of $u_{mf}$ due to vibration is also evident and even more pronounced for high moisture contents than for FCC catalyst. At high vibration intensities, $u_{mf}$ values approach the same level, regardless of the moisture content. The cohesive forces added by moisture seem to be counteracted completely by vibration. In contrast to FCC catalyst, this point is not yet reached at $\Lambda = 1.21$. All of the above observations may be explained by the lower specific surface area of the Cellets, compared to FCC catalyst. Thus, less water is located inside the pores of the Cellets and more water is present at the particles' surfaces, resulting in more pronounced cohesive effects due to liquid bridges between Cellets.
McLaren's correlation is able to predict $u_{mf}$ of Cellets, given the moisture content is low, i.e. corresponding to the second drying period. Hence, the particle surfaces are not covered with water and no additional cohesive effects, in form of liquid bridges, are

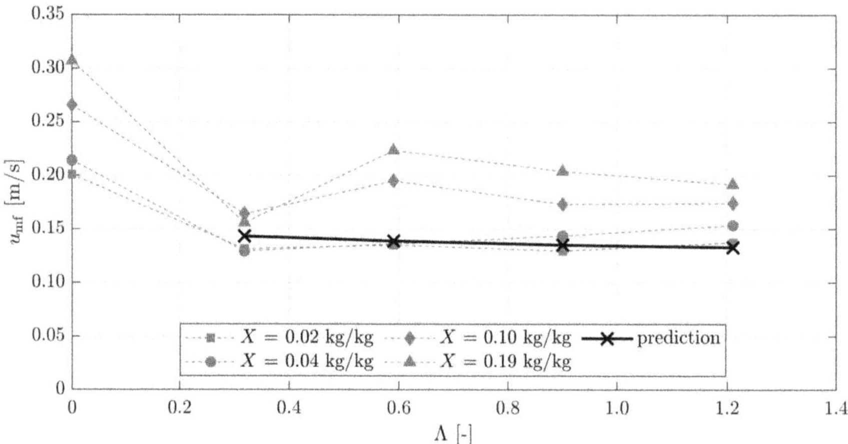


Figure 3.16: Measured minimum fluidization velocity ($u_{mf}$) of Cellets against vibration intensity ($\Lambda$), showing influence of particle moisture content ($X$) and predictions after McLaren et al. (2021).

present. This confirms the validity of McLaren's correlation for vibrated fluidized beds of Geldart B particles, with the limitation to dry particles. Additionally, the observed effect of resonance frequency of the bed material is not considered in the correlation.

Overall, the observed trends for Cellets and FCC catalyst are expected. Quantification of $u_{mf}$ in dependence of moisture content and vibration intensity is used in the modeling of fluidized beds hydrodynamic and drying kinetics (see chapters 4 and 5). Investigation of hydrodynamics of whole milk powder (WMP) in analogy to FCC catalyst and Cellets is significantly more complex, due to the WMP's solubility in water. As introduced in chapter 3.2, the stickiness of WMP poses additional challenges to the investigation of $u_{mf}$. The humidified WMP does not show classical fixed bed behavior, i.e. linear decrease of pressure drop with decreasing gas velocity. Instead it is characterized by channeling, which results in pressure drop curves, deviating from the required fixed bed regime for determination of $u_{mf}$. An exemplary case is shown in Figure 2.1. Hence, only limited data after the conventional determination of $u_{mf}$ is available for dry WMP. These results are shown in Figure 3.17 and compared to the correlation by McLaren et al. (2021). The data show decreasing $u_{mf}$ with increasing vibration intensity. The predicted $u_{mf}$ lie below the measured values. This is to be expected, due to the cohesive nature of WMP, which is not considered in the correlation for $u_{mf}$ (McLaren et al., 2021). However, sufficient characterization of WMP under the influence of vibration and moisture content cannot be made based on the available $u_{mf}$ data. Consequently, the lower limit of fluidization of WMP is characterized in terms of complete fluidization velocity $u_{cf}$ in the following chapter.

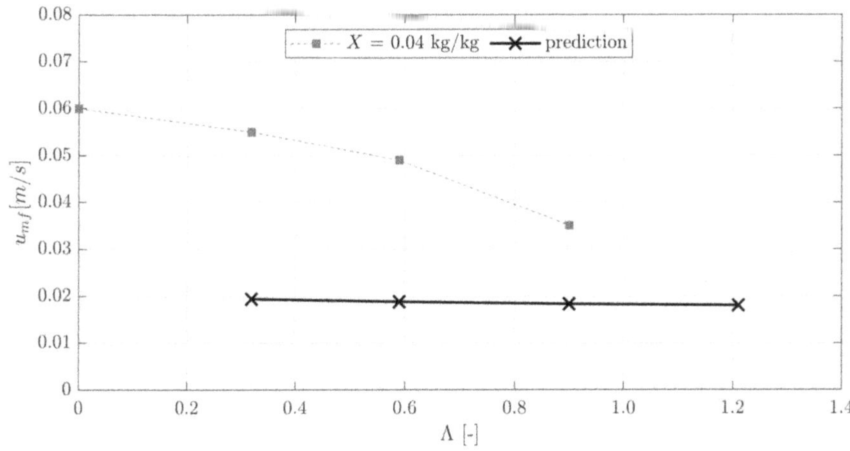


Figure 3.17: Measured minimum fluidization velocity ($u_{\mathrm{mf}}$) of whole milk powder (WMP) against vibration intensity ($\Lambda$), showing available data for dry milk powder ($X = 0.04\,\mathrm{kg\,kg^{-1}}$) and predictions after McLaren et al. (2021).

### 3.3.4 Complete Fluidization Velocity

The complete fluidization velocity $u_{\mathrm{cf}}$ of whole milk powder is determined based on pressure drop measurements, as depicted in Figure 2.1. Varied parameters are powder moisture content ($0.04\,\mathrm{kg\,kg^{-1}}$, $0.06\,\mathrm{kg\,kg^{-1}}$ and $0.09\,\mathrm{kg\,kg^{-1}}$) as well as initial bed mass (12 kg and 25 kg, corresponding to a fixed bed height of approximately 19 cm and 38 cm, respectively). Some parameter settings are investigated repeatedly (a, b) to confirm reproducibility. The results are shown in Figure 3.18. The data from repeated experiments show very similar results. Thus, it is concluded that the results are reproducible. Furthermore, it is observed that variation of the bed mass within the investigated range has no impact on the lower limit of fluidization. Resonance frequency according to equation 2.2 is 1.8 Hz and 2.6 Hz, for 25 kg and 12 kg bed mass, respectively. Thus, resonance effects are not evident in the investigated cases of WMP.

In analogy to the previously discussed investigations, the moisture content has significant influence on $u_{\mathrm{cf}}$. An increase of moisture content from $0.04\,\mathrm{kg\,kg^{-1}}$ to $0.06\,\mathrm{kg\,kg^{-1}}$ results in an increase of $u_{\mathrm{cf}}$ by roughly a factor of three. Elevation of moisture content from $0.04\,\mathrm{kg\,kg^{-1}}$ to $0.09\,\mathrm{kg\,kg^{-1}}$ causes an increase of $u_{\mathrm{cf}}$ by a factor of six. This is evident for all investigated vibration intensities. Furthermore, vibration results in decrease of $u_{\mathrm{cf}}$, as expected and observed for $u_{\mathrm{mf}}$ of the insoluble particles in chapter 3.3.3. In fact, $u_{\mathrm{cf}}$ decreases more rapidly for higher vibration intensities. The highest vibration intensity of 1.21 leads to a reduction of $u_{\mathrm{cf}}$ by a factor of two, compared to the non-vibrated case. This is shown for all investigated moisture levels. The determined $u_{\mathrm{cf}}$ values mark the lowest possible gas velocities for the operation of vibrated fluidized beds at room temperature. At elevated temperatures, the stickiness curve of milk powder

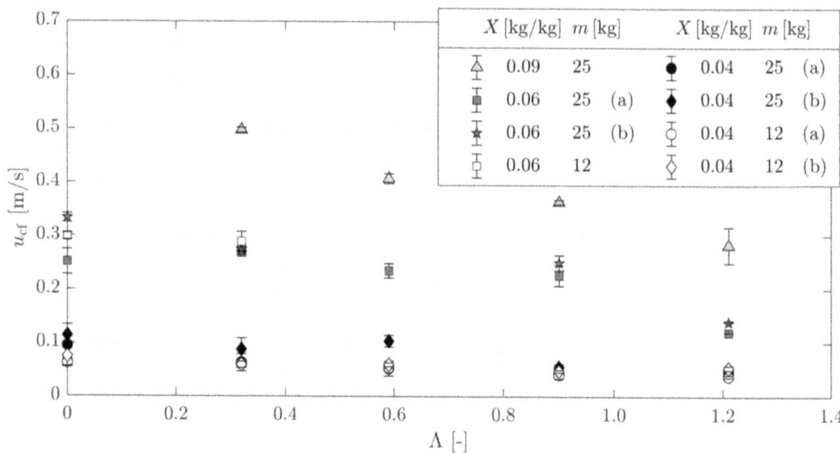


Figure 3.18: Measured complete fluidization velocity ($u_{cf}$) of whole milk powder (WMP) against vibration intensity ($\Lambda$); showing influence of particle moisture content ($X$), bed mass ($M_P$) and repeated investigations (a, b) for confirmation of reproducibility.

needs to be considered as well (Palzer, 2005).

### 3.3.5 Hydrodynamics and Bubble Characteristics

The goal of this investigation is the development of a semi-empirical model, describing the hydrodynamics of cohesive powders in vibrated and non-vibrated fluidized beds. Thus, whole milk powder (WMP) is investigated, in terms of bubble volume ratio and bed expansion. This is analogous to widely applied correlations used in the modeling of fluidized beds (Alaathar et al., 2013, Hilligardt and Werther, 1986). Since fluidized beds of cohesive powders are not only penetrated by bubbles but also by horizontal cracks (Mawatari et al., 2015), the determination of diameters of individual bubbles or equivalent bubble diameters can be misleading when the overall bubble volume fraction is of interest, e.g. in the flow sheet simulation of VFB processes (Alaathar et al., 2013). Thus, the bubble volume fraction is used for investigation of bubble characteristics. In order to account for the entire fraction of the gas without direct contact with particles, all solid free parts (bubbles and cracks, regardless of shape and size) are considered in the bubble volume fraction.

Hydrodynamics are investigated locally via capacitance probe and globally via high speed camera imaging. Validation of the applied methods is discussed here, results and model development are discussed in chapter 4.1.

### Validation

WMP particles constantly stick to the capacitance probe's electrodes, making local investigation of bubble characteristics impossible and leaving optical investigations via high speed camera as the only viable option. Validation of accuracy and limits of said optical investigations is required, specifically the transferability of data gathered at the wall to the inside of a 3D bed. Thus, non-cohesive glass beads with similar particle size and the same Geldart group are used. Therein, capacitance measurements are conducted in positions close to the glass wall and in the middle of the bed at gas velocities, identical to the investigations with WMP. High speed camera images are taken simultaneously with capacitance measurements at the wall. The capacitance probe can only be operated in non-vibrated fluidized beds, thus only the non-vibrated case is available for validation of image analysis data.

For comparison of capacitance measurements and image analysis, the area of the image analysis is reduced to the area around the needles of the capacitance probe. The results of this calibration are shown in Figure 3.19. The expected increase in bubble volume fraction with increasing gas velocity is measured by both techniques for excess gas velocities ($u - u_{\mathrm{mf}}$) of up to $0.3\,\mathrm{m\,s^{-1}}$. For higher velocities, the results from capacitance measurements begin to increase stronger than the bubble volume fraction measured via image analysis. In fact, the bubble volume fraction derived via image analysis stays constant for excess gas velocities higher than $0.3\,\mathrm{m\,s^{-1}}$. This observation is to be expected, since bubbles move towards and coalesce in the center of the bed, when the gas velocity is increased. Thus, upwards movement of particles takes place mainly in the center and a downwards movement of particles occurs in the vicinity of the walls. This flow pattern is called core-annulus flow (Wiesendorf and Werther, 2000).

In contrast to the image analysis method, the capacitance probe cannot measure directly at the wall. Hence, the image analysis will result in lower bubble volume fractions for high gas velocities, where the core annulus flow is present. For excess gas velocities lower than $0.25\,\mathrm{m\,s^{-1}}$, it appears that the core-annulus flow is not developed and thus capacitance measurements at both positions as well as image analysis give approximately the same results. The threshold value of $c_{P,bub}$ is identified to be 0.12.

Consequently, it is assumed that the image analysis method yields reasonable results for WMP if the superficial gas velocity is low enough, so that the core annulus flow is not yet developed. For glass beads, this is valid for excess gas velocities lower than $0.25\,\mathrm{m\,s^{-1}}$. Due to the cohesiveness of WMP, the needles of the capacitance probes are permanently covered with powder, which prevents measurement of bubble characteristics in WMP with capacitance probes. Thus, the only possible option is digital image analysis of high speed images taken through the glass wall of the VFB. Results of the image analysis are presented in chapter 4.1. An additional condition for investigation of WMP via image analysis is that WMP particles must not stick to the glass wall and block the view inside the fluidized bed. This limits the available data to investigations with dry WMP. At

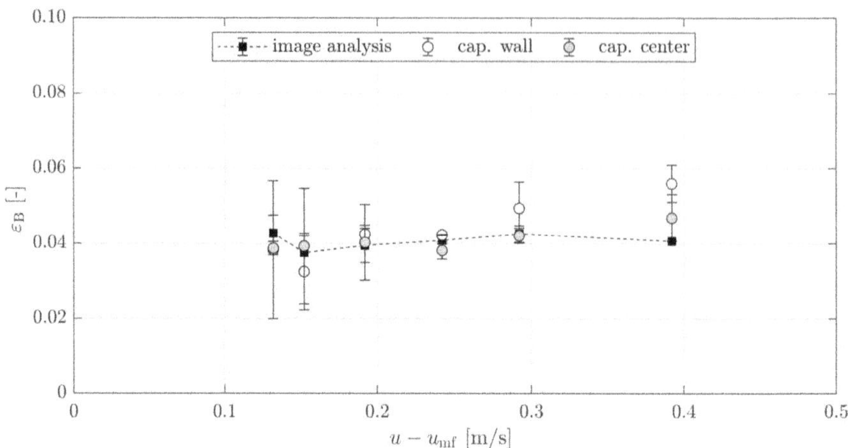


Figure 3.19: Comparison of bubble volume fraction ($\varepsilon_B$), measured via image analysis and capacitance probe at wall (cap. wall) and center (cap. center) of the VFB, plotted against excess gas velocity ($u - u_{mf}$).

increased moisture content, stickiness increases and particle adherence to the glass wall makes image analysis impossible.

### 3.3.6 Residence Time Distribution

Magnetizable tracer particles are used for investigation of residence time characteristics. Cellet500 particles are coated with magnetic paint (*Alpina Magneto*) in a batch fluidized bed. The production process was developed and described in detail by Pietsch et al. (2018). The same tracer particles, used in this work, were previously confirmed to match all requirements (see 2.1.4) and used in investigations by Pietsch (2018), Pietsch et al. (2020). Insignificant wear and deterioration of the tracer particles was confirmed (Pietsch et al., 2020).

The VFB dryer is operated in the continuous setup. Particle mass flow and moisture content are monitored to confirm steady-state operation of the dryer, before tracer particles are introduced. The tracer particles are added to the feed material stream via a three-way-valve as a pulse signal. The three-way-valve is located between the rotary valve and the inlet into the drying chamber. The entire particle mass flow is collected at the outlet in sampling containers. The containers are changed in defined time intervals and the contained mass of particles is measured. Tracer particles are separated from the regular particles with a magnetic rod. The quality of the separation is controlled visually, based on the white color of regular Cellets and dark gray color of tracer particles. Then, the mass of tracer particles in every sample is measured and residence time distribution (RTD) curves are plotted. The density distribution of the RTD is derived via:

$$E(t_\mathrm{i}) = \frac{\dot{M}_\mathrm{T,i}}{M_\mathrm{T,total}}\,. \tag{3.8}$$

In the course of the experiments, small deviations in particle mass flow rate are unavoidable, due to the periodically intermittent feed of the rotary valve, small changes in gas flow rate and periodic control of the heaters. In order to account for these small fluctuations, the mass flow of tracer particles is calculated by normalizing the mass of tracer in every sample by the total mass of the corresponding sample and multiplying this with the average particle mass flow over the course of the experiment $\overline{\dot{M}_\mathrm{P}}$. Thus, the corrected tracer mass flow rate $\dot{M}_\mathrm{T,i}$ in equation 3.8 results from:

$$\dot{M}_\mathrm{T,i} = M_\mathrm{T,i}\frac{\overline{\dot{M}_\mathrm{P}}}{M_\mathrm{sample,i}}\,. \tag{3.9}$$

The total mass of injected tracer particles $M_\mathrm{T,total}$ corresponds to approximately 2 % of the bed mass. As a fraction of the tracer particles stays in the bed for several hours, the experiments are stopped after about an hour. In analogy to Pietsch (2018), the remaining data points are determined via two-parameter Weibull extrapolation, defined in equation 3.10. Recovery of the majority of the tracer particles during the experimental time is guaranteed by ensuring that the expected hydrodynamic residence time is well exceeded by the duration of the experiment:

$$E_\mathrm{Weibull}(t) = \omega_1\,\omega_2\,(\omega_1\,t)^{\omega_2-1}\,\exp\left(-\,(\omega_1\,t)^{\omega_2}\right) \quad \text{for:}\, t, \omega_1, \omega_2 > 0. \tag{3.10}$$

A double logarithmic Weibull mesh is used to fit the Weibull scale factor $\omega_1$ and the form factor $\omega_2$:

$$\ln\left(-\ln\left(1 - F(t)\right)\right) = \omega_2 \ln(\omega_1) + \omega_2 \ln(t). \tag{3.11}$$

Plotting the left hand side of equation 3.11 against $\ln(t)$ results in a straight, with slope $\omega_2$ and the intercept $j$. The Weibull scale factor $\omega_1$ is calculated by:

$$\omega_1 = e^{j/\omega_2}\,. \tag{3.12}$$

The analysis parameters and corresponding equations have been introduced in chapter 2.1.4, integrals are calculated using the trapezoidal rule in *MATLAB*. Parameter studies are conducted to elucidate the influence of vibration on the residence time in dependence of gas flow rate, gas temperature, particle mass flow rate, bed height and particle moisture content. The results of the RTD investigations with Cellets are discussed in chapter 4.2.

Additionally, FCC catalyst tracer particles have been produced and tested according to the procedures introduced by Pietsch et al. (2018). However, cohesive forces between regular FCC particles and FCC tracer are too high to achieve proper separation with the magnetic rod. Alternative measurement techniques might enable the RTD analysis of FCC catalyst. Possible approaches for future research could be capacitance measurements in the outgoing mass flow or detection of induction current, resulting from the magnetizable tracers.

## 3.4 Lab-Scale Fluidized Bed Dryer

To allow for investigation of different dryer size and geometry, a lab-scale fluidized bed *GF3* (*Glatt GmbH*, Germany) is used for investigation of drying kinetics under batch operation and performance of continuous drying experiments. The respective details and configurations are introduced in the following chapters.
General features of the *GF3* are the conical fluidization chamber with a diameter of 180 mm at the distributor plate and 250 mm at the top (320 mm above the distributor). An expansion zone is located above the fluidization chamber to allow for deceleration and return of entrained particles. A cylindrical part with internal filters for removal of fines from the gas stream is located above. The distributor consists of a perforated plate (for mechanical support of the bed material) and a sieve mesh.
The *GF3* is operated with ambient air, conveyed by a suction fan. The air flow rate is regulated by adjustment of the rotational speed via frequency converter. The volumetric flow rate of the process air is measured with a vane anemometer. Temperature and relative humidity of the process air are measured at the inlet and the outlet. Furthermore, temperature probes (PT100) are located in the windbox and inside the bed. Air temperature is adjusted by an electrical heater. Control of process parameters as well as data logging is done via an SPS software. A flow sheet and a photograph of the *GF3* lab-scale fluidized bed (for batch operation) are shown in figure 3.20. The bed height during operation is measured through an observation window in the drying chamber.

### 3.4.1 Batch Drying

For batch drying experiments, the *GF3* is equipped with an inline capacitance moisture analyzer *SampIn* (*AVA*, Germany), allowing for measurements of particle moisture content with high time resolution. One measurement is conducted approximately every ten seconds and all measurements over a period of two minutes are averaged, resulting in one data point used for analysis. Additionally, samples can be taken from the bed during operation via a separate port.
Material specific drying curves are determined by batch drying experiments for Cellets and $\gamma$-$Al_2O_3$ particles in the *GF3* dryer. The results are discussed in chapter 4.3.1.

### 3.4.2 Continuous Drying

Continuous fluidized bed drying is conducted in the *GF3* dryer for Cellets and $\gamma$-$Al_2O_3$ particles. A simplified flowchart of the *GF3* dryer in continuous operation is shown in Figure 3.21. The wet feed is stored in a feed bunker and transported into the dryer,

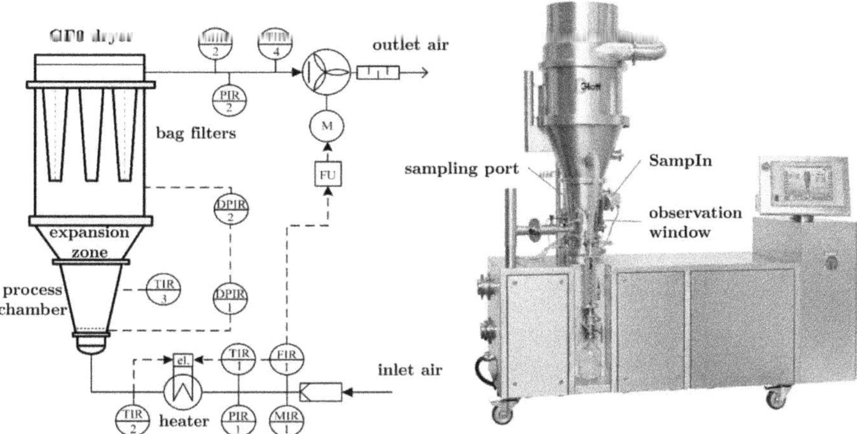


Figure 3.20: Photograph (right) and simplified flowchart of the *GF3* fluidized bed (*Glatt GmbH*, Germany) in batch operation (left); modified from Alaathar (2017). Showing key measurement locations of temperature (TIR), humidity (MIR) and flow rate (FIR) of air, as well as bed temperature (TIR 3). During batch drying, particle moisture content is measured inline via *SampIn* probe.

using a vibrating chute. Particles leave the drying chamber through an outflow hole. The height of the outflow hole is varied between 7 cm and 12 cm above the distributor, allowing for investigation of the influence of bed height on the drying process. Mass flow of the feed is controlled by adjusting the vibration intensity of the vibrating chute.

Moisture content of the dried particles is measured externally via thermo-gravimetric analysis. Parameter studies for continuous, steady-state drying are conducted to generate extensive data for validation of the flowsheet simulation model. The results are discussed in detail in chapter 5.

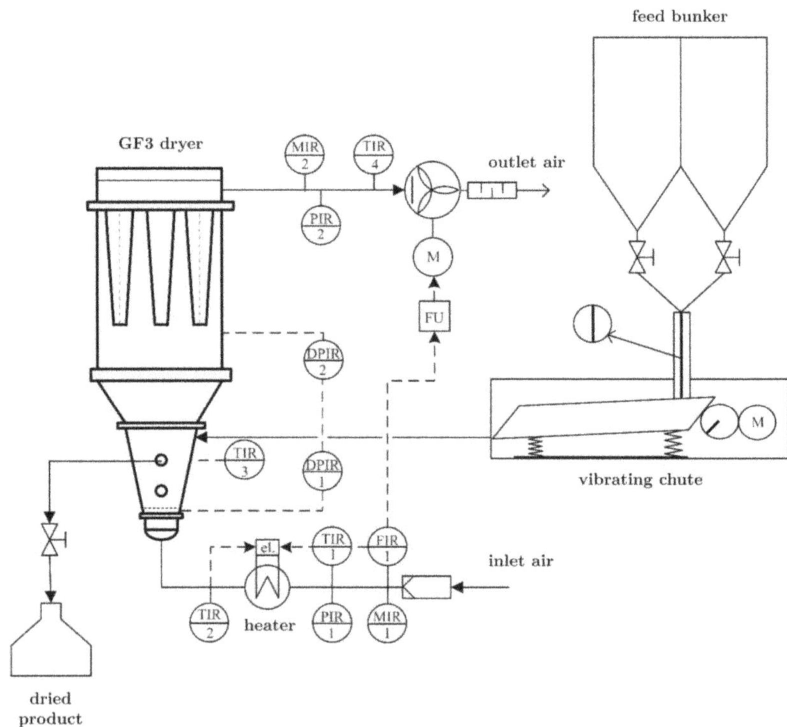


Figure 3.21: Simplified flowchart of the *GF3* fluidized bed (*Glatt GmbH*, Germany) in continuous drying operation, modified from Alaathar (2017). Showing key measurement locations of temperature (TIR), humidity (MIR) and flow rate (FIR) of drying air, as well as bed temperature (TIR 3). In continuous operation, wet feed is stored in a bunker and transported into the dryer via vibrating chute. Dried particles exit the dryer through an outflow hole. The height of the outflow hole can be adjusted between 7 cm and 12 cm, to investigate the influence of the bed height on the process.

# 4

# Model Development

Development of a comprehensive flowsheet simulation model for vibrated and non-vibrated fluidized bed drying is the goal of this thesis. The hydrodynamics of Geldart B and D particles have been investigated extensively in the past and applied successfully in flowsheet simulation models (Alaathar, 2017, Burgschweiger, 2000, Hilligardt, 1986). The robustness and accuracy of the hydrodynamics model by Hilligardt (1986) has been confirmed by several researchers (Abbasi et al., 2019, Bakshi et al., 2017, Burgschweiger, 2000, Haus et al., 2018, Kramp et al., 2012). Thus, hydrodynamics of non-cohesive particles (Geldart groups A, B and D) are modeled according to Hilligardt (1986) in the scope of this work.
However, there is a lack of hydrodynamic models for fine and cohesive particles (Geldart group C) and cohesive particles of Geldart group A, such as whole milk powder. Furthermore, many hydrodynamic models lack the influence of vibration. Thus, a hydrodynamic model for vibrated and non-vibrated fluidized beds of fine and cohesive powders is developed in chapter 4.1.
Furthermore, the influence of vibration on particle residence time distribution is investigated and included in the developed model (see chapter 4.2). Modeling of drying kinetics is covered in chapter 4.3.

## 4.1 Hydrodynamics of Cohesive Powders under Vibration

Whole milk powder (WMP) is investigated in terms of bubble volume ratio, bed expansion and $u_{\mathrm{mf}}$. This is analogous to widely applied correlations used in the modeling of fluidized beds (Alaathar et al., 2013, Hilligardt and Werther, 1986). The measurement of hydrodynamic characteristics, namely bed porosity and bubble volume fraction, are described in detail in chapter 3.3.5. The results were published earlier in Lehmann et al. (2019).

### 4.1.1 Bubble Volume Fraction

As introduced in chapter 3.3.5, the bubble volume fraction is measured based on high speed camera images. Time and spatial averages are calculated from the image series (100 fps for 5 s). The cohesive nature of the investigated WMP significantly complicates the collection of usable data. Parts of the VFB dryer's glass wall are often covered with milk powder, preventing the detection and investigation of bubbles. This occurs for experiments with and without vibration alike. The cohesiveness increases further with increasing powder moisture content. Consequently, only limited data for the lowest investigated moisture content ($0.04\,\mathrm{kg\,kg^{-1}}$) are available.
The measured bubble volume fraction is plotted against vibration intensity for a gas velocity of $0.2\,\mathrm{m\,s^{-1}}$ in Figure 4.1. Since $u_{\mathrm{mf}}$ changes with vibration, the excess gas velocity varies between $0.14\,\mathrm{m\,s^{-1}}$ in the non-vibrated case and $0.165\,\mathrm{m\,s^{-1}}$ for the highest vibration intensity. These values lie below the identified maximum excess gas velocity ($0.25\,\mathrm{m\,s^{-1}}$), ensuring the validity of image analysis at the wall to represent conditions inside the 3D bed (see chapter 3.3.5). Slightly decreasing bubble volume fraction is evident for increasing vibration intensity. Cano-Pleite et al. (2016) observed the same trend. Following these observations, it may be concluded that the bubble volume fraction is reduced by vertical vibration of the fluidized bed.
Using the presented experimental data, a model description of bubble volume fraction is developed. The model is based on original correlations for the bubble volume fraction $\varepsilon_{\mathrm{B}}$ from Hilligardt and Werther (1986):

$$\varepsilon_{\mathrm{B}} = \frac{\dot{V}_{\mathrm{B}}}{u_{\mathrm{B}}}. \tag{4.1}$$

The correlation for the visible bubble flow rate $\dot{V}_{\mathrm{B}}$ is expanded to account for the impact of vibration, using the vibration intensity $\Lambda$ and the measured values for $u_{\mathrm{mf}}$ at the respective vibration parameters in the following equation:

$$\dot{V}_{\mathrm{B}} = \psi \cdot (u - u_{\mathrm{mf}}) \cdot \frac{1}{1+\Lambda}. \tag{4.2}$$

Therein, the bubble rise velocity $u_{\mathrm{B}}$ is calculated, using the original correlation by Hilligardt and Werther (1986) including the hydrodynamic parameter $\psi$ for Geldart group A (see Table 4.3), the gravitational acceleration $g$ and the bubble rise velocity $u_{\mathrm{B}}$:

$$u_{\mathrm{B}} = \dot{V}_{\mathrm{B}} + 0.71 \cdot \Theta\sqrt{g \cdot d_{\mathrm{B}}}. \tag{4.3}$$

For the cohesive WMP, $u_{\mathrm{B}}$ is calculated according to the correlation by Zou et al. (2011), which was specifically developed for fine and cohesive particles:

$$d_{\mathrm{B}} = 0.21 \frac{(u - u_{\mathrm{mb}})^{0.49}\left(z + 4\sqrt{A_0}\right)^{0.48}}{g^{0.2}}, \tag{4.4}$$

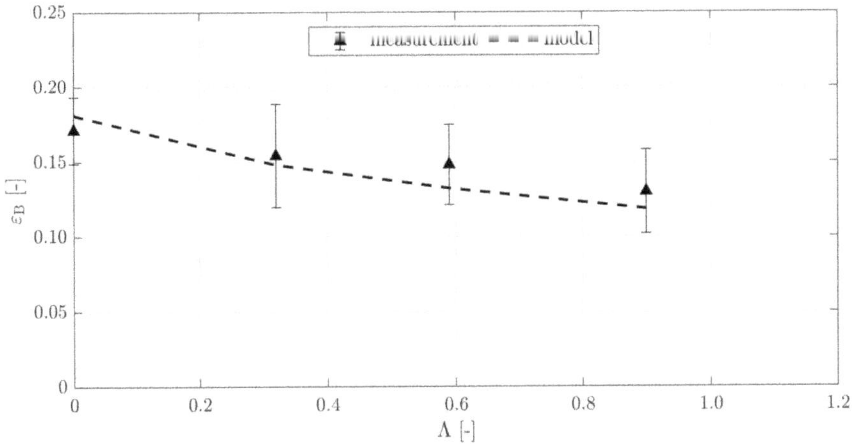


Figure 4.1: Comparison of measured and predicted bubble volume fraction ($\varepsilon_B$) against vibration intensity ($\Lambda$) of whole milk powder (WMP) with moisture content $X = 0.04\,\mathrm{kg\,kg^{-1}}$ at gas velocity $u = 0.2\,\mathrm{m\,s^{-1}}$ .

with $u_{mb}$ being the minimum bubbling velocity, $z$ the height above the distributor and $A_0$ the cross-sectional area of one opening of the distributor. When porous or sintered plate distributors are used, $A_0$ becomes very small and may be set to zero (Zou et al., 2011). Investigations of this work revealed that $u_{mb} \approx u_{mf}$. Thus, $u_{mf}$ is used in equation 4.4 instead of $u_{mb}$.

Figure 4.1 shows predicted bubble volume fraction in dependence of vibration intensity. The decrease of bubble volume fraction with increasing vibration intensity is predicted by the model. All model predictions lie well within the standard deviation through each series of images, depicted by error bars.

### 4.1.2 Bed Porosity

The bed porosity is an important parameter in the design and operation of fluidized beds. It includes the bubble volume fraction and is directly linked to the operational bed height (see equation 3.5). The measured bed expansion of WMP for varying vibration intensities is plotted against superficial gas velocity in Figure 4.2.

It shows a strong increase in bed porosity around the point of minimum fluidization, which directly correlates to expansion of the bed. The onset of the observed bed expansion shifts towards lower gas velocities with increasing vibration intensity. This is in accordance with the reduction of the onset of fluidization, i.e. the reduction of minimum as well as complete fluidization velocity, discussed in chapters 3.3.3 and 3.3.4. Furthermore, the data in Figure 4.2 show strong increase in overall bed expansion with increasing vibration intensity.

The finding of increased bed expansion under vibration is particularly interesting in

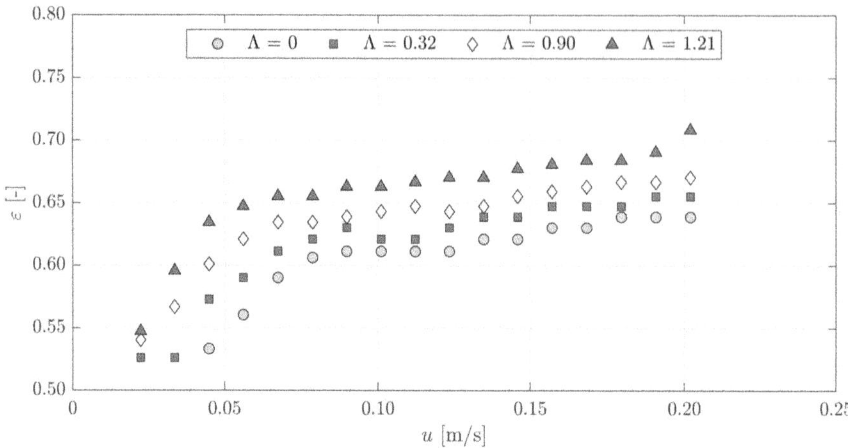


Figure 4.2: Measured bed porosity ($\varepsilon$) of whole milk powder (WMP) with moisture content $X = 0.04\,\mathrm{kg\,kg^{-1}}$ against gas velocity ($u$), showing influence of vibration intensity ($\Lambda$).

combination with the simultaneous reduction of bubble volume fraction. Consequently, vibration forces a portion of gas into the suspension phase that is part of the bubble phase in non-vibrated systems. This explains reported enhancement of drying kinetics in vibrated fluidized beds (Cruz et al., 2005), considering the widely accepted assumption of direct heat and mass transfer between particle and gas only taking place in the suspension phase. The bed porosity $\varepsilon$ is modeled according to Alaathar (2017) with the following equation:

$$\varepsilon = 1 - (1 - \varepsilon_{\mathrm{B}})\,(1 - \varepsilon_{\mathrm{S}})\,, \tag{4.5}$$

using the bubble volume fraction $\varepsilon_{\mathrm{B}}$ and porosity of the suspension phase $\varepsilon_{\mathrm{S}}$. In order to account for the influence of vibration, $\varepsilon_{\mathrm{B}}$ is derived after equation 4.1. For $\varepsilon_{\mathrm{S}}$, a modified version of the correlation by Richardson and Zaki (1954) is used:

$$\varepsilon_{\mathrm{S}} = \varepsilon_{\mathrm{mf}} \cdot \left(\frac{u_{\mathrm{S}}}{u_{\mathrm{mf}}}\right)^{1/(4.65\cdot(1+\Lambda))}\,, \tag{4.6}$$

where $\varepsilon_{\mathrm{mf}}$ is the bed porosity at minimum fluidization and $u_{\mathrm{S}}$ the velocity of the suspension gas, which is calculated according to the correlation by Clift et al. (1983):

$$u_{\mathrm{S}} = u_{\mathrm{mf}}\,(1 + 1.5\varepsilon_{\mathrm{B}})^{2/3}\,. \tag{4.7}$$

Figure 4.3 compares model predictions and measurements of bed porosity of dry WMP ($X = 0.04\,\mathrm{kg\,kg^{-1}}$) at varied gas velocities and vibration intensities. It is evident that the predictions of the developed model represent the influence of vibration of bed porosity accurately over the investigated range of gas velocities. Deviations between measure-

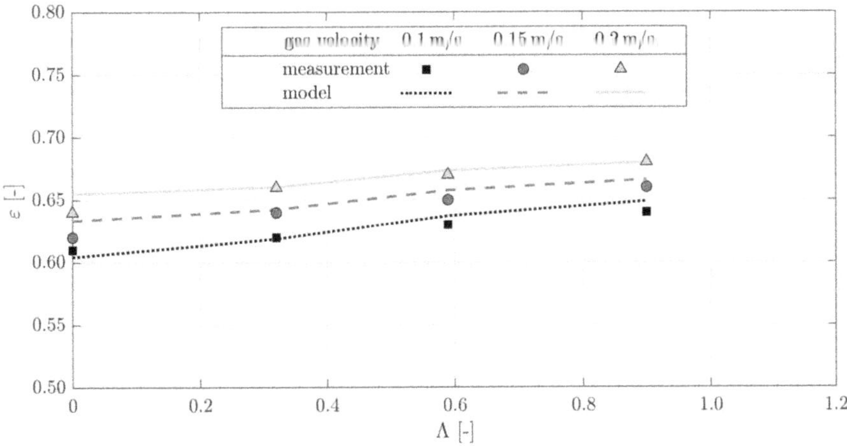


Figure 4.3: Comparison of measured and predicted bed porosity ($\varepsilon$) against vibration intensity ($\Lambda$) of whole milk powder (WMP) with moisture content $X = 0.04\,\mathrm{kg\,kg^{-1}}$; showing influence of gas velocity ($u$).

ments and predictions are only observed in the non-vibrated cases.

The proposed correlations for the bubble volume fraction (equations 4.1 through 4.4) and the bed porosity (equations 4.5 through 4.7) for cohesive particles of Geldart groups A and C are designed to account for the impact of vibration. They are based on established correlations and include the vibration intensity in a way that they equal the original correlations, if no vibration is applied (i.e. $\Lambda = 0$). In addition to the vibration intensity, minimum fluidization velocity and bed porosity at minimum fluidization are affected by vibration. Measured values of said parameters are used in the prediction of this work, due to the lack of reliable predictive models for these parameters.

## 4.2 Residence Time Distribution

A total of 25 experiments is conducted and presented to investigate the residence time characteristics of the VFB dryer, using Cellets as bed material. The residence time distribution (RTD) is investigated by comparison of density distribution curves or E-curves $E(t)$ and calculation of characteristic values, such as mean residence time $\tilde{\tau}$ and theoretical number of tanks $K$, according to the tank in series (TIS) model (see chapter 2.1.4). The RTD is investigated with respect to the influence of bed height, particle mass flow, vibration and particle moisture content. First, the reproducibility of the experimental procedure and results is investigated. If not stated otherwise, the particle moisture content is kept constant at $X = 0.040 \pm 0.005\,\mathrm{kg\,kg^{-1}}$.

### 4.2.1 Reproducibility

Experiments with selected parameter settings are conducted threefold, in order to asses the accuracy and reproducibility of the experimental procedure and the resulting RTD data. The parameter settings, as well as resulting hydrodynamic mean residence time, measured mean residence time and tank number are listed in Table 4.1. The outflow height is 8 cm and the target particle mass flow is set to 200 g min$^{-1}$. The measured average particle mass flow does not vary by more than 5 % and is thereby considered close enough to the target value to allow for comparison. The results are illustrated in Figure 4.4, showing reproducibility tests without vibration (top) and two different vibration intensities (middle and bottom). The presented density curves of the RTD show almost perfect overlap for repeated investigations. Furthermore, the averaged mean residence time and average theoretical tank number are calculated for every parameter set. The respective standard deviations do not exceed 3.3 % for average residence time and 5.6 % for the theoretical tank number. Based on these results, it is concluded that the applied procedures deliver reproducible results. Thus, and due to the long duration of experiments (between 8 and 10 hours per experiment) further experiments are only conducted once.

Another observation made in Table 4.1 is that hydrodynamic mean residence times are approximately 10 % to 15 % larger than measured mean residence times in all cases. These large deviations may be explained by dead zones in the bed, i.e. areas in the fluidized bed, where particles stay much longer compared to the average of the particle mass flow. Such dead zones are likely located in the corners of rectangular fluidized bed, as friction between particles and wall is stronger and hinders movement of particles to other parts of the bed.

A second aspect that may increase the formation of dead zones is the design of the outflow hole. Its diameter only equals approximately 10 % of the width of the wall. Thus, particles are likely to be pushed towards the wall with the outflow hole by newly added particles from the opposite side, but without directly leaving the dryer.

The use overflow weirs, spanning the entire width of the dryer, is likely to reduce such dead zones and minimize the difference between hydrodynamic and measured mean residence time. For constructional reasons and mechanical stability, this is not an option for the VFB dryer, used in this thesis. However, this observation should be considered in the design of continuous fluidized beds. It is further of interest for investigations regarding fluidized beds with several compartments (see for example (Bachmann et al., 2017, Diez et al., 2019)) and in particular for the design of particle transfer geometry between the compartments, as investigated by Pietsch (2018).

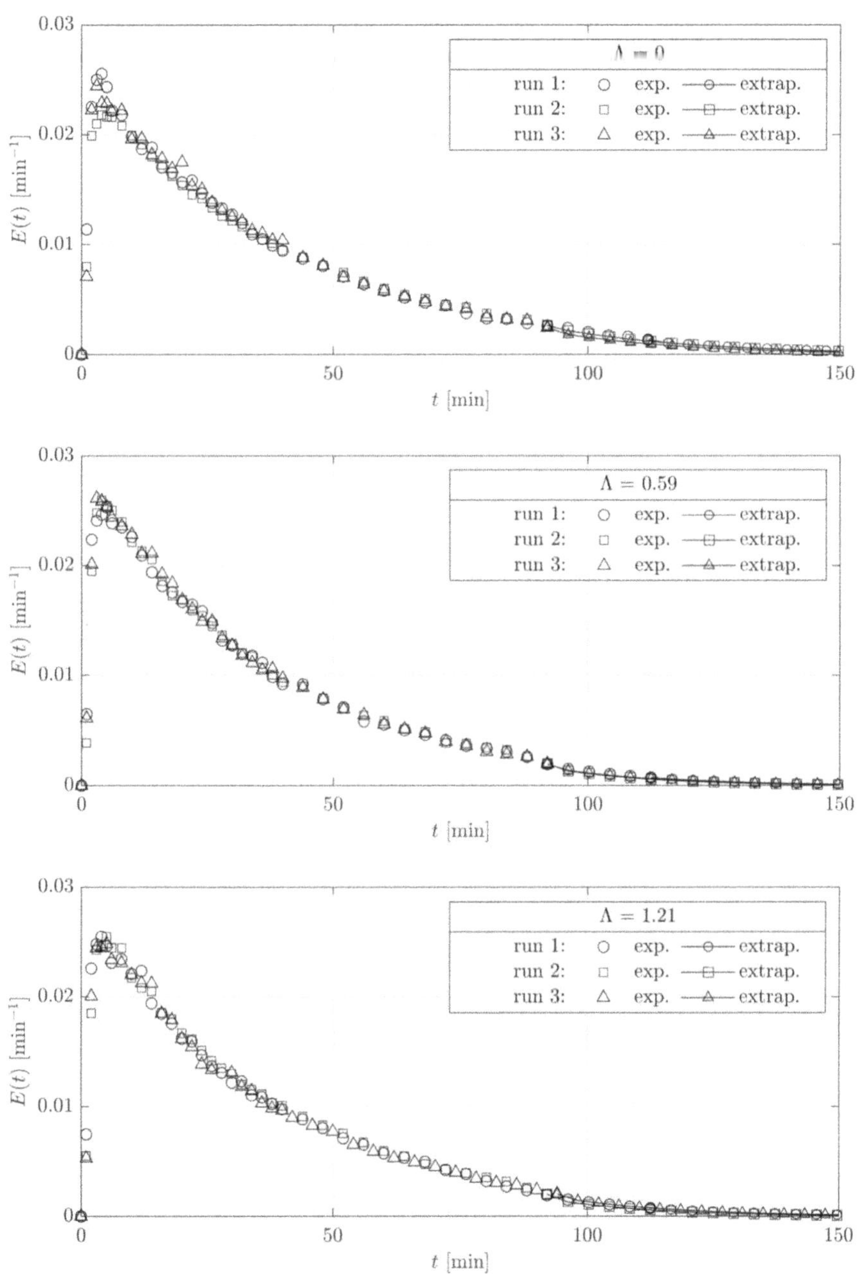


Figure 4.4: Residence time distribution density curves ($E(t)$) against time ($t$); showing repeated experiments with identical configurations (outlet height $H_{\text{out}} = 8\,\text{cm}$ and target mass flow $\dot{M}_{\text{P,targ}} = 200\,\text{g}\,\text{min}^{-1}$) for different vibration intensities ($\Lambda$). Also showing measured data points (exp.) and extrapolated data points (extrap.). Corresponding parameters, resulting mean residence time and theoretical tank numbers are listed in Table 4.2.

Table 4.1: Summary of parameter settings and residence time characteristics from reproducibility experiments with constant outflow height of 8 cm and particle moisture content of 0.04 kg kg$^{-1}$ (corresponding RTD density curves depicted in Fig. 4.4).

| rep. [-] | $\Lambda$ [-] | $\dot{M}_P$ [g min$^{-1}$] | $\bar{\tau}$ [min] | averaged $\bar{\tau}$ [min] | $\tilde{\tau}$ [min] | averaged $\tilde{\tau}$ [min] | $K$ [-] | averaged $K$ [-] |
|---|---|---|---|---|---|---|---|---|
| 1 | 0 | 202.88 | 39.93 | | 36.39 | | 1.23 | |
| 2 | 0 | 198.58 | 40.37 | $39.71 \pm 0.64$ | 36.30 | $35.67 \pm 1.17$ | 1.26 | $1.27 \pm 0.039$ |
| 3 | 0 | 210.31 | 38.85 | | 34.32 | | 1.31 | |
| 1 | 0.59 | 201.91 | 38.33 | | 32.15 | | 1.32 | |
| 2 | 0.59 | 207.26 | 36.57 | $36.90 \pm 1.07$ | 31.60 | $31.78 \pm 0.32$ | 1.39 | $1.35 \pm 0.032$ |
| 3 | 0.59 | 210.17 | 35.78 | | 31.60 | | 1.35 | |
| 1 | 1.21 | 206.80 | 37.14 | | 32.59 | | 1.33 | |
| 2 | 1.21 | 209.18 | 37.05 | $37.10 \pm 0.04$ | 32.43 | $32.54 \pm 0.09$ | 1.45 | $1.36 \pm 0.076$ |
| 3 | 1.21 | 206.34 | 37.12 | | 32.60 | | 1.30 | |

$\bar{\tau}$: hydrodynamic mean residence time; $\tilde{\tau}$: measured mean residence time

### 4.2.2 Influence of Particle Mass Flow and Bed Height

The influence of particle mass flow and bed height is investigated in this chapter. The latter is controlled via varied height of the outflow. The resulting RTD density curves are shown in Figure 4.5 for cases without vibration (top) and with vibration (bottom). The corresponding characteristic values are listed in Table 4.2 (top section).
The results show that the combination of highest particle mass flow and lowest bed height leads to the narrowest RTD and shows the lowest mean residence time. In contrast, the combination of highest bed height and lowest mass flow results in the widest RTD and highest mean residence time, also showing a theoretical tank number of 1.16 which is the closest to ideal CSTR of all experiments. Furthermore, it is observed that an increase in bed height by factor two leads to increased mean residence time by approximately factor 1.6. On the other hand, increase in particle mass flow by factor two reduces mean residence time by half. As a consequence, the RTD density curves for the combinations 'high bed height, high mass flow' and 'low bed height, low mass flow' are almost identical. These observed trends and dependencies agree with previously reported trends in literature (see chapter 2.1.4). The quantified results are used in this work for flowsheet simulation of the VFB dryer.

### 4.2.3 Influence of Moisture Content

The results of RTD investigations regarding the influence of moisture content are listed in Table 4.2 (middle section) and illustrated in Figure 4.6 for non-vibrated fluidization (top) and vibration (bottom). The experimental results show that particle moisture content has an impact on the RTD characteristics. Reduced mean residence times as well as narrower RTDs are observed with increasing particle moisture content. Thereby, the theoretical tank numbers increase, representing reduced dispersion of particles. The decreased dispersion due to increased moisture content is more pronounced at higher particle mass flow. Both observations confirm previous literature findings (see chapter 2.1.4).
In particular, comparison of particle moisture content of $0.04\,kg\,kg^{-1}$ and $0.07\,kg\,kg^{-1}$ shows only small increase in theoretical tank number. In contrast, cases with particle moisture content of $0.18\,kg\,kg^{-1}$ show significant increase in theoretical tank number. This observation is based on the fact that the particles at $0.04\,kg\,kg^{-1}$ and $0.07\,kg\,kg^{-1}$ are in the second drying period, whereas for higher moisture content, they are in the first drying period. In the first drying period, the entire particle surface is covered with water, resulting in high degree of liquid bridges, hindering dispersion of particles. During the second drying period, the particle surface is mainly or completely deprived of fluid, which lowers the cohesiveness and supports dispersion of particles. The described observations are evident in cases with vibration (Figure 4.6 bottom) and cases without vibration (Figure 4.6 top). The influence of vibration on RTD characteristics is discussed comprehensively in the next chapter.

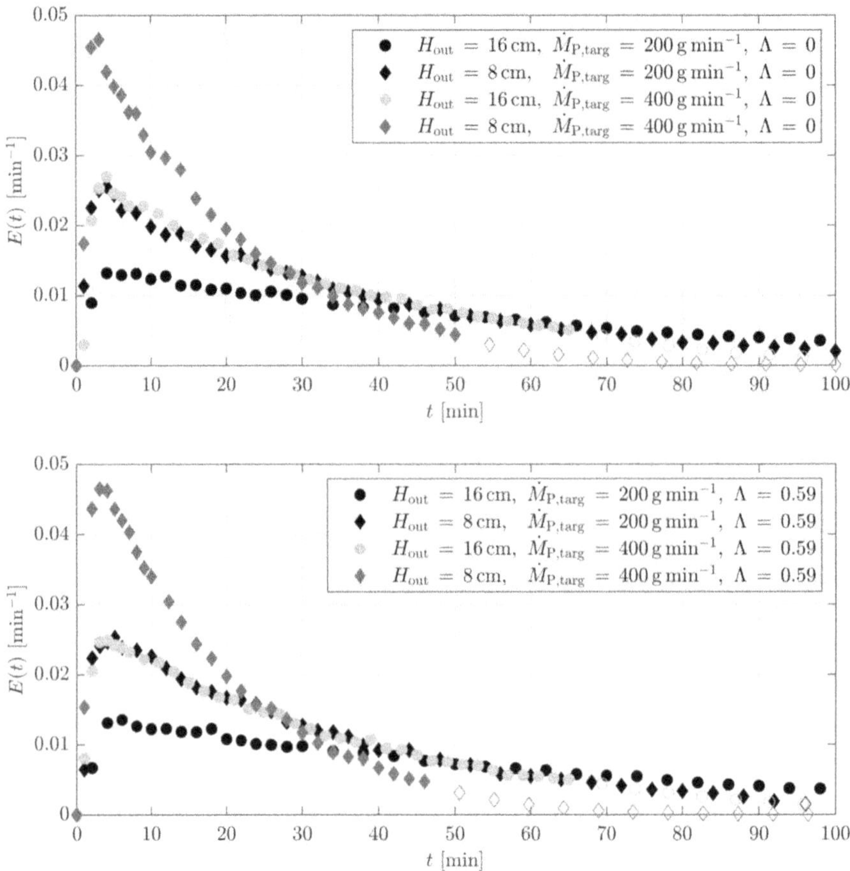


Figure 4.5: Residence time distribution density curves ($E(t)$) against time ($t$); showing influence of outlet height ($H_{\text{out}}$), target particle mass flow ($\dot{M}_{\text{P,targ}}$) for non-vibrated fluidization (top) and vibration (bottom). Solid markers indicate measured data points, hollow markers represent extrapolated data. Corresponding parameters, resulting mean residence time and theoretical tank numbers are listed in Table 4.2 (top section).

Table 4.2: Summary of residence time characteristics from parameter studies regarding particle mass flow ($\dot{M}_P$), outflow height ($H_{out}$), vibration intensity ($\Lambda$) and particle moisture content ($X$); also indicating the corresponding graphical illustrations of residence time distribution density curves.

| corresponding Fig. | $\Lambda$ [-] | $H_{out}$ [cm] | $X$ [kg kg$^{-1}$] | $\dot{M}_P$ [g min$^{-1}$] | $\bar{\tau}$ [min] | $K$ [-] |
|---|---|---|---|---|---|---|
| Fig. 4.5 (top) | 0 | 16 | 0.04 | 203.99 | 63.25 | 1.16 |
| | 0 | 8 | 0.04 | 202.88 | 36.39 | 1.23 |
| | 0 | 16 | 0.04 | 397.35 | 29.64 | 1.33 |
| | 0 | 8 | 0.04 | 401.08 | 18.88 | 1.32 |
| Fig. 4.5 (bottom) | 0.59 | 16 | 0.04 | 212.28 | 60.87 | 1.23 |
| | 0.59 | 8 | 0.04 | 201.91 | 32.15 | 1.32 |
| | 0.59 | 16 | 0.04 | 403.28 | 31.29 | 1.27 |
| | 0.59 | 8 | 0.04 | 392.69 | 17.17 | 1.35 |
| Fig. 4.6 (top) | 0 | 8 | 0.04 | 198.58 | 36.30 | 1.26 |
| | 0 | 8 | 0.07 | 188.44 | 34.74 | 1.24 |
| | 0 | 8 | 0.18 | 213.86 | 32.40 | 1.32 |
| | 0 | 8 | 0.04 | 401.08 | 18.88 | 1.32 |
| | 0 | 8 | 0.18 | 418.78 | 15.92 | 1.44 |
| Fig. 4.6 (bottom) | 0.59 | 8 | 0.04 | 201.91 | 32.15 | 1.32 |
| | 0.59 | 8 | 0.07 | 201.87 | 31.67 | 1.28 |
| | 0.59 | 8 | 0.18 | 207.77 | 30.15 | 1.42 |
| | 0.59 | 8 | 0.04 | 392.69 | 17.17 | 1.35 |
| | 0.59 | 8 | 0.18 | 409.50 | 14.18 | 1.67 |
| Fig. 4.7 (a) | 0 | 8 | 0.04 | 198.58 | 36.30 | 1.26 |
| | 0.59 | 8 | 0.04 | 201.91 | 32.15 | 1.32 |
| | 1.21 | 8 | 0.04 | 206.80 | 32.59 | 1.33 |
| Fig. 4.7 (b) | 0 | 8 | 0.04 | 401.08 | 18.88 | 1.32 |
| | 0.59 | 8 | 0.04 | 392.69 | 17.17 | 1.35 |
| | 1.21 | 8 | 0.04 | 398.76 | 16.74 | 1.40 |
| Fig. 4.7 (c) | 0 | 16 | 0.04 | 203.99 | 63.25 | 1.16 |
| | 0.59 | 16 | 0.04 | 206.45 | 61.85 | 1.19 |
| | 1.21 | 16 | 0.04 | 212.28 | 60.87 | 1.23 |
| Fig. 4.7 (d) | 0 | 16 | 0.04 | 397.35 | 29.64 | 1.33 |
| | 0.59 | 16 | 0.04 | 403.28 | 31.29 | 1.27 |
| | 1.21 | 16 | 0.04 | 396.09 | 31.09 | 1.38 |

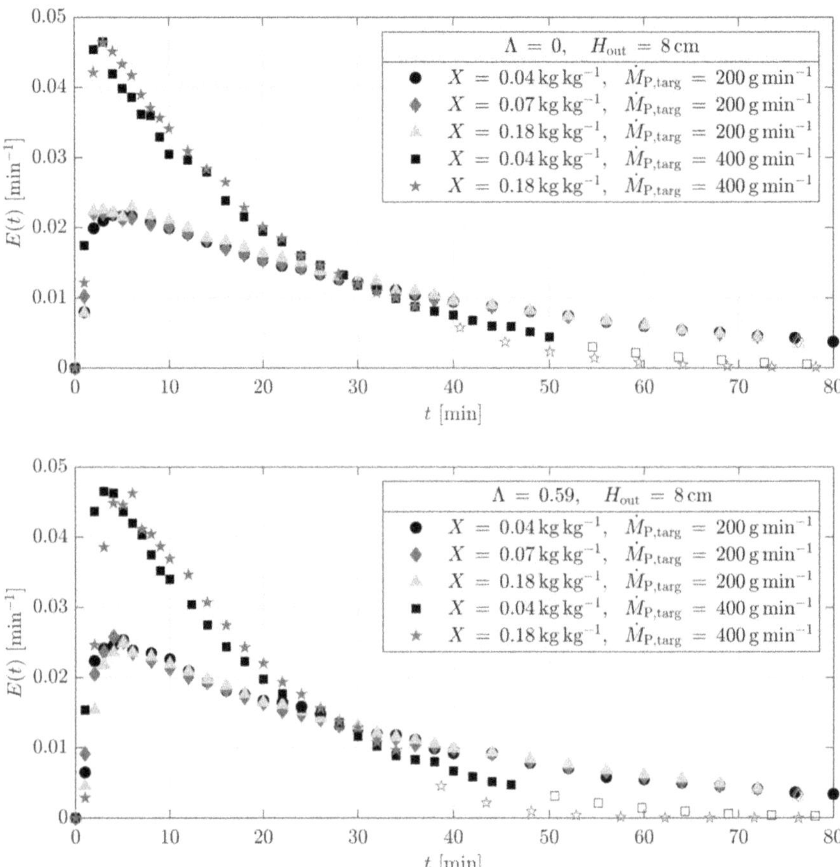


Figure 4.6: Residence time distribution density curves ($E(t)$) against time ($t$); showing influence of particle moisture content ($X$) and target particle mass flow ($\dot{M}_{\mathrm{P,targ}}$) for non-vibrated case (top) and with vibration (bottom). Solid markers indicate measured data points, hollow markers represent extrapolated data. Corresponding parameters, resulting mean residence time and theoretical tank numbers are listed in Table 4.2 (middle section).

### 4.2.4 Influence of Vibration

Besides the influence of moisture content, also the impact of vibration is investigated in the experiments, shown in Figure 4.6. Considering the characteristic values listed in Table 4.2 (middle section), vibration leads to lowered mean residence times and increased theoretical tank numbers, compared to non-vibrated cases. Thus, it may be concluded that the reduced dispersion is caused by addition of vibration. This is in agreement with the previously discussed observations regarding the impact of vibration. It also counteracts the expected increase of residence time due to slightly reduced particle mass flow rates, evident in some of the cases (see e.g. cases with high target particle mass flow rates and $X = 0.18\,\mathrm{kg\,kg^{-1}}$ in the middle section of Table 4.2).

This is particularly interesting, considering the few previous studies regarding the impact of vibration (Brod et al., 2004, Han et al., 1991, Satija and Zucker, 1986). In all of those cases, the vibration was found to reduce dispersion of particles and thus, increase the theoretical tank number. This effect was attributed to the horizontal component of the applied vibration, transporting the particles towards the outlet of the dryer. In contrast to that, strictly vertical vibration was applied in the this work's experiments. Consequently, other phenomena are likely involved when it comes to the impact of vibration on RTD characteristics.

Further parameter studies regarding the impact of vibration are illustrated in Figure 4.7 and the corresponding mean residence times and theoretical tank numbers are listed in Table 4.2 (bottom section).

The RTD density curves for low bed height with low and high particle mass flow are shown in Figure 4.7 (a) and (b), respectively. The RTD density curves for non-vibrated cases deviate slightly from vibrated cases. For both investigated vibration intensities, the RTD curves are almost identical. Thus, different vibration intensities have no obvious effect on particle RTD in the investigated range. As observed before, vibration results in lower mean residence time and increased theoretical tank number, meaning decreased dispersion of particles (see Table 4.2). The observed effect of vibration is more pronounced for the cases with low mass flow rate. Higher mass flow rates also reduce dispersion. When the bed height is doubled, no impact of vibration is evident in the RTD density curves (Figure 4.7 (c) and (d)). Neither, theoretical tank numbers nor mean residence time are affected by vibration.

These observations may be explained by the design of the dryer and the interaction between particle bed and distributor under vibration. The vertical vibration is transferred to the particles via vertical movement of the distributor. On its way upwards in the vibration cycle, particles are subject to additional upwards momentum from the distributor. This additional momentum (on top of the upwards moving gas flow and movement of bubbles as well as eruption of bubble at the top of the bed) results in increased bed expansion and thereby supports movement of particles through the outflow hole. Increased bed expansion due to vibration was observed in literature (Satija and

Zucker, 1986) and is discussed in chapter 4.1.
For increased bed height however, the extra momentum of particles at the distributor is damped by the increased mass of particles. Consequently, increased bed height counteracts the additional expansion of the bed due to vibration.

Over all investigated parameter combinations, the theoretical tank numbers vary between 1.16 and 1.67, which is close to the ideal CSTR behavior. However, the tank number and the resulting particle residence time distributions have significant impact on the drying process. Hence, the measured tank numbers, listed in Table 4.1 and Table 4.2, are used in the flowsheet simulations of the VFB dryer to model the RTD density curves for the respective process parameters, according to equations 2.12 and 2.15.

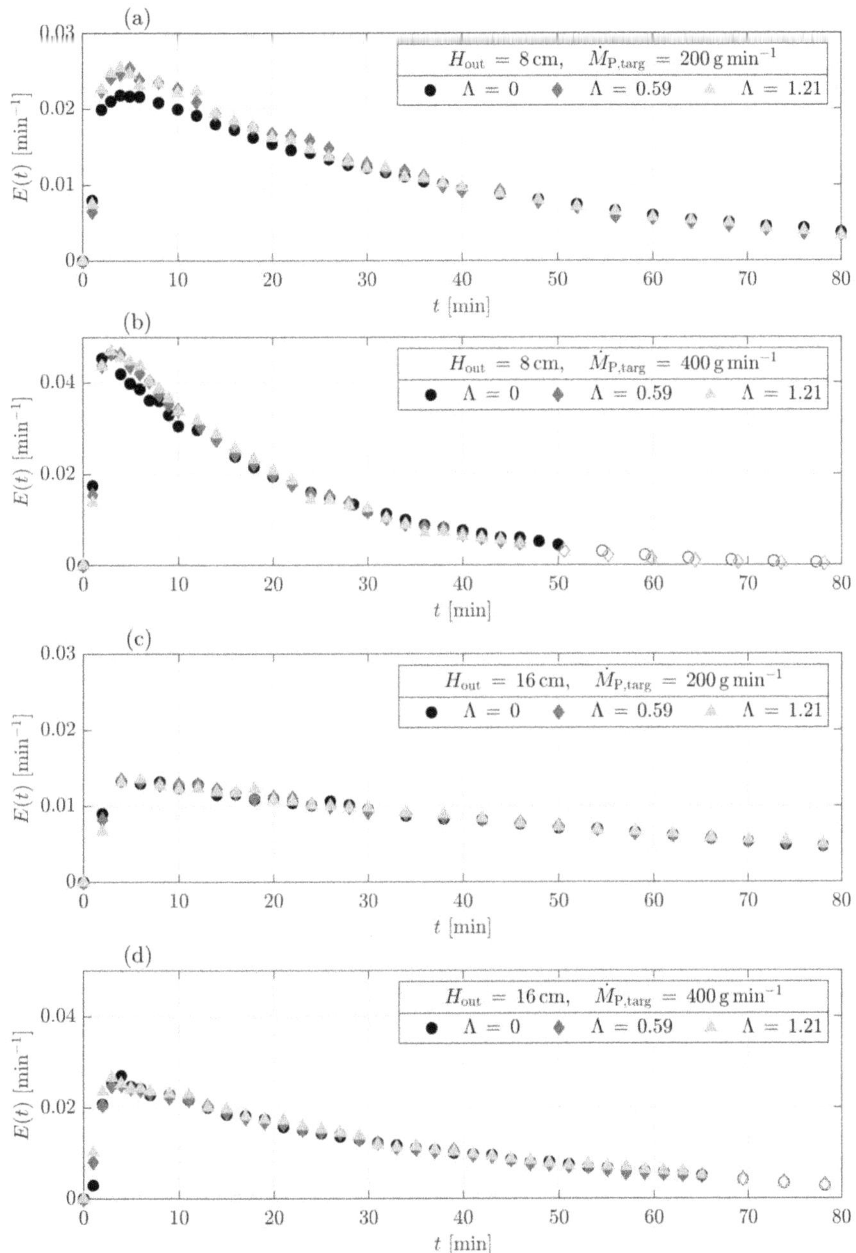


Figure 4.7: Residence time distribution density curves ($E(t)$) against time ($t$); showing influence of vibration intensity ($\Lambda$) for varied combinations of outlet height ($H_{\text{out}}$) and target particle mass flow ($\dot{M}_{\text{P,targ}}$). Solid markers indicate measured data points, hollow markers represent extrapolated data. Corresponding parameters, resulting mean residence time and theoretical tank numbers are listed in Table 4.2.

## 4.3 Drying Kinetics

As previously discussed in chapter 2.2, the best way of representing the drying kinetics in fluidized bed drying models is utilization of material specific drying curves. Therein, material specific resistances towards drying in the second drying period are summarized in one function. The Normalized Characteristic Drying Curve (NCDC) is the most commonly used approach. However, a significant weakness of the NCDC is its dependency on the critical moisture content $X_{cr}$ and thus, dependence on the measurement conditions. The Reaction Engineering Approach (REA) also expresses the material specific drying kinetics in a single curve, but does not require knowledge of $X_{cr}$. The REA has been applied to single droplet and film drying (Chen, 2008). Applicability of the REA to model fluidized bed drying processes with and without vibration (as introduced in chapter 2.2.2) is topic of the following chapter. The results were previously published (Lehmann et al., 2020).

### 4.3.1 Applicability of the Reaction Engineering Approach on Fluidized Bed Drying

Batch drying experiments with $\gamma$-$Al_2O_3$ particles are conducted in the *GF3* dryer, as described in chapter 3.4.1. Measured drying kinetics, i.e. particle moisture content against time, are shown in Figure 4.8 with respect to the influence of drying temperature and gas velocity at constant bed mass. As expected, increasing gas temperature and gas velocity leads to decreasing drying time. Figure 4.9 constitutes the influence of bed mass on the drying kinetics. An increase in bed mass leads to increased drying time. In all experiments, the equilibrium moisture content of the particles $X_{eq}$ is $(0.015 \pm 0.001)$ kg/kg. The critical moisture content $X_{cr}$ is $(0.075 \pm 0.005)$ kg/kg in all cases. Thus, $X_{eq}$ and $X_{cr}$ are considered constant in the scope of the discussed investigations.
Threefold repetition of the parameter settings (60 °C, $u = 4 \cdot u_{mf}$, 3 kg bed mass) is conducted to confirm reproducibility of experimental results. Drying kinetics data from the reproducibility runs are shown in Figure 4.10. Deviations between individual experiments, conducted at different days, lie within $\pm 2$ % of the average moisture content for any given data point. Hence, reproducibility of the experimental procedure and results is confirmed and the remaining parameter combinations are only investigated once.
The REA curve of $\gamma$-$Al_2O_3$ is determined, based on the presented data after equation 2.43. Respective vapor pressures are calculated based on relative humidity measurements of the process air before and after the dryer. Mass balance analysis showed that certain amounts of inleaked air entered the expansion zone and led to reduced moisture content measurements at the outlet of the dryer. Alaathar (2017) made the same observation during his experiments with the same setup. In the first drying period, complete saturation of air leaving the bed may be assumed. The inleaked air has ambient conditions. The mass flow of inleaked air is calculated via mass balance and the humidity measurement is corrected for each experiment, based on the deviation between saturation and

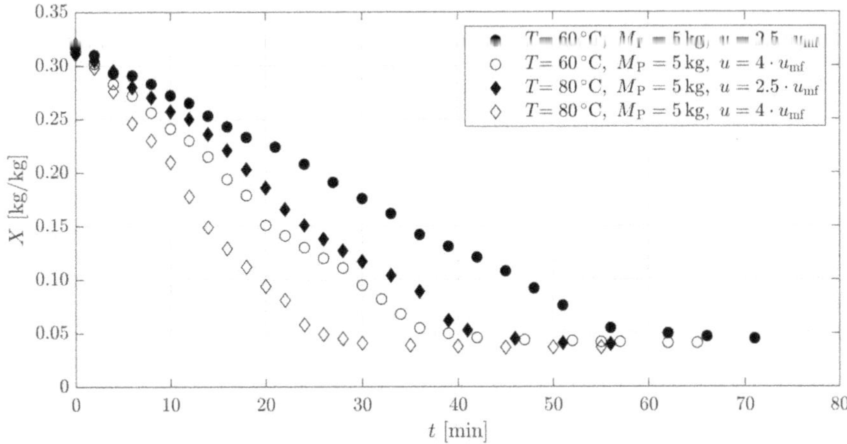


Figure 4.8: Moisture content of $\gamma$-$Al_2O_3$ particles against drying time ($t$), showing influence of gas velocity ($u$) and drying temperature ($T$) on the drying kinetics in the *GF3* dryer.

measured values.

Figure 4.11 shows the REA curves for all drying experiments in form of normalized activation energy against excess moisture content ($X - X_{eq}$). All curves, determined at differing parameter combinations, collapse in one material specific curve. As a key feature of the REA, this is expected and confirms the successful expression of material specific drying resistance of $\gamma$-$Al_2O_3$, independent of process parameters. An exponential regression curve is fitted through the data, using the method of least-non-linear-squares with least-absolute-residuals criteria. The equation and the determination coefficient $R^2$ are given in Figure 4.11. Detection of one material specific curve is confirmed by an $R^2$ value of larger than 99 %. The curves have the expected shape, reaching equilibrium state towards the end of the drying (unity) and a material specific resistance to evaporation of zero in the first drying period.

The determined REA curve for the $\gamma$-$Al_2O_3$ is compared to NCDC data from Burgschweiger et al. (1999) to confirm the accuracy of the REA curve. Burgschweiger et al. (1999) measured NCDCs for single particles in two different setups: a micro balance and a drying channel. They reported significant deviations between the resulting NCDC, based on different drying conditions, leading to differences in $X_{cr}$. Hence, the influence of drying conditions is still reflected in the NCDC despite the normalization, which should result in one material specific function (Burgschweiger et al., 1999).

NCDCs are usually plotted against the normalized moisture content. Thus, the determined REA data for $\gamma$-$Al_2O_3$ are plotted versus the normalized moisture content. As before, exponential regression is performed and the resulting equation is depicted in Figure 4.12. By definition, the NCDC has a constant value of 1 in the first drying period

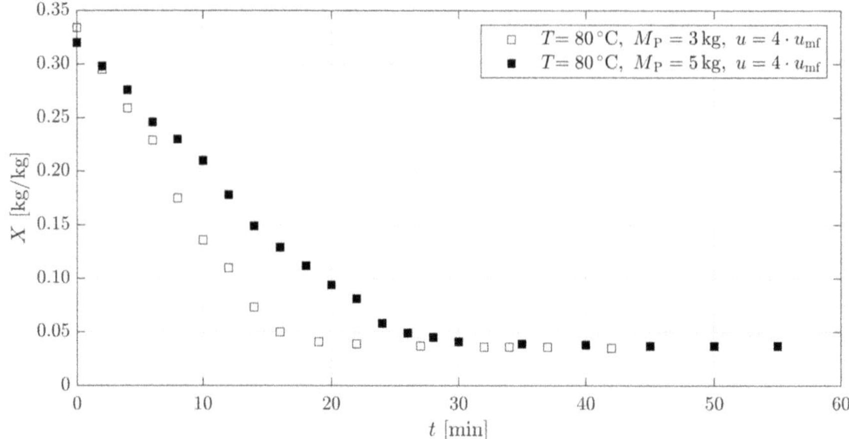


Figure 4.9: Moisture content of $\gamma$-$Al_2O_3$ particles against drying time ($t$), showing influence of bed mass ($M_P$) on the drying kinetics in the *GF3* dryer.

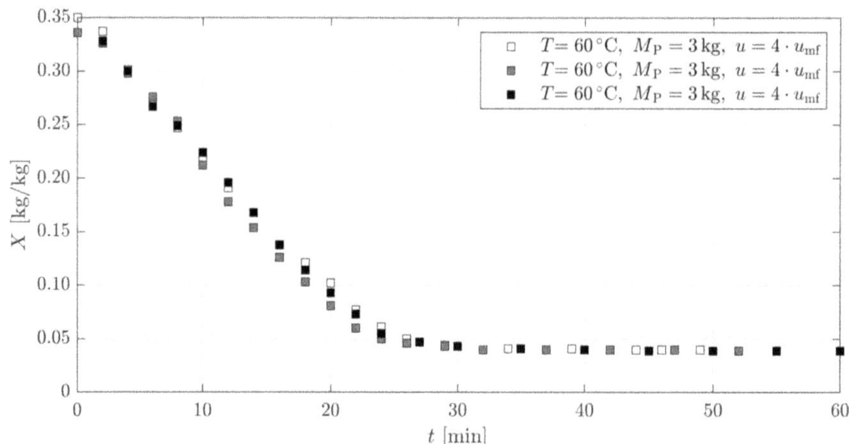


Figure 4.10: Moisture content of $\gamma$-$Al_2O_3$ particles against drying time ($t$), showing repeated experiments and identical process parameter in the *GF3* dryer.

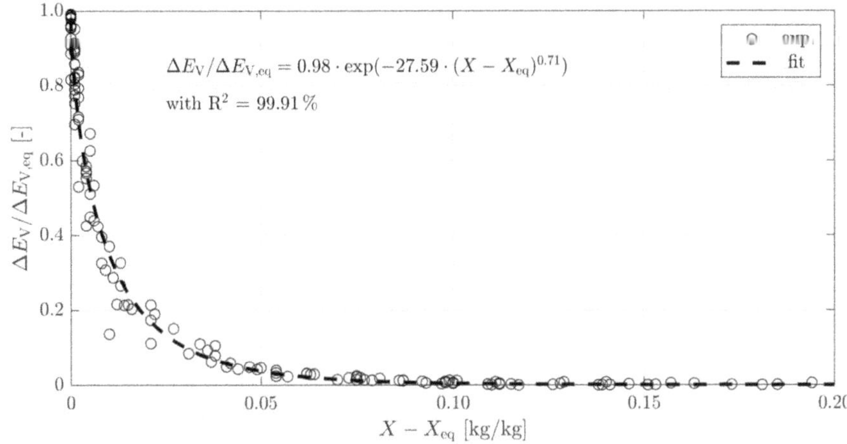


Figure 4.11: Reaction Engineering Approach data (normalized activation energy $\Delta E_V/\Delta E_{V,eq}$) against moisture content ($X - X_{eq}$) for all experiments with $\gamma$-$Al_2O_3$ particles, including material specific regression function.

and the origin represents the equilibrium state. The REA curve is inverted, i.e. subtracted from unity according to equation 2.45, allowing for direct comparison of REA and NCDC.

Figure 4.13 shows the comparison of the inverted REA curve and NCDC literature data for the same material from (Burgschweiger et al., 1999). The REA curve coincides well with the data from the drying channel. In particular for low moisture contents the REA curve lies in between Burgschweiger's curves. Towards the end of the first drying period (moisture content approaching $X_{cr}$), the REA deviates slightly from NCDC data and does not reach unity at $X_{cr}$. The reason lies in the different determination methods. Transition from first to second drying period is a precise point by definition of the NCDC. This transition is not easily identified as a precise point, even for single particle investigations, as $X_{cr}$ is influenced by a number of factors (Burgschweiger et al., 1999). The transition from first to second drying period is a range of moisture content rather than a sharp point in fluidized beds, resulting from distributed properties, such as particle size, temperature or moisture content. Furthermore, the NCDC uses two different mathematical descriptions with a designed change at the critical moisture content. In contrast, the REA is one continuous mathematical description of the drying process. Based on this, small deviation between the REA curve and the NCDC values around $X_{cr}$ are inevitable.

Nonetheless, high similarity between NCDC and REA of $\gamma$-$Al_2O_3$ is evident. This proves that material specific drying curves may be measured independently of $X_{cr}$ and the underlying uncertainties in the measurement of $X_{cr}$. Additionally, fluidized bed drying experiments and the REA are adequate for determination of material specific drying

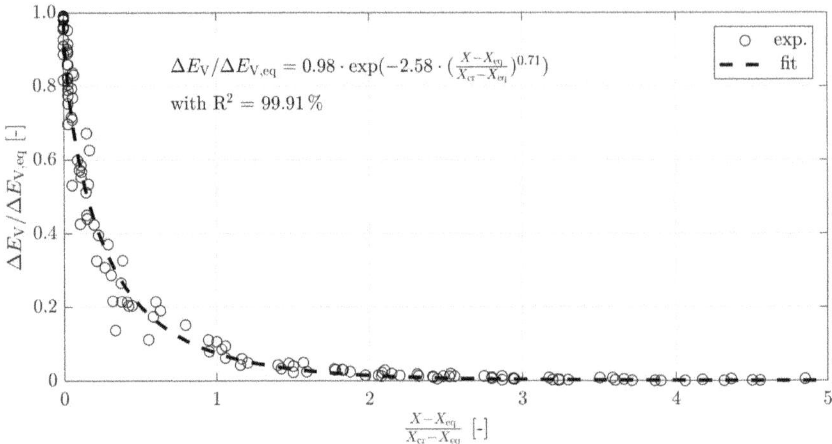


Figure 4.12: Normalized activation energy ($\Delta E_V/\Delta E_{V,eq}$) against normalized moisture content; showing Reaction Engineering Approach data for all experiments with $\gamma$-$Al_2O_3$ particles, including material specific regression function.

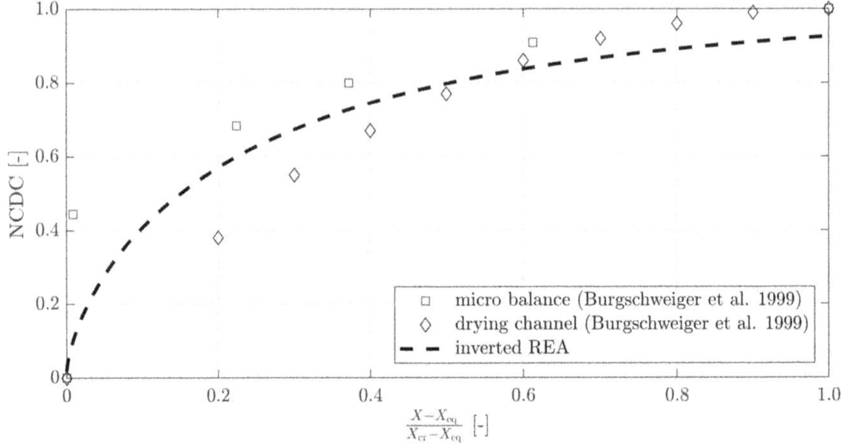


Figure 4.13: Inverted Reaction Engineering Approach curve against normalized moisture content of $\gamma$-$Al_2O_3$ particles in comparison to NCDC data by Burgschweiger et al. (1999).

curves. This is of particular value, when materials are to be characterized that have particle sizes too small for single particle experiments.

It is concluded that the REA data derived in this work are suitable to describe the material specific drying kinetics in fluidized bed drying. This is based on the high similarity between the REA curve and the NCDC data from (Burgschweiger et al., 1999), which has been successfully used in the modeling of fluidized bed drying applications (Alaathar, 2017, Alaathar et al., 2013, Burgschweiger and Tsotsas, 2002, Chen et al., 2017). Consequently, the REA data is used in the modeling of vibrated fluidized bed drying in the scope of this work.

### 4.3.2 Reaction Engineering Approach Curves for Modeling of Fluidized Bed Drying

Material specific drying curves are determined according to the REA for Cellets and FCC catalyst in analogy to $\gamma$-$Al_2O_3$ particles. Batch fluidized drying experiments are performed with Cellets in the *GF3* dryer. FCC catalyst is investigated in the VFB dryer, as introduced in chapter 3.3.2.

Drying kinetics of Cellets are measured to determine the material specific drying curve. The drying kinetics show the same trends as for $\gamma$-$Al_2O_3$ particles and are illustrated in Figure B.1 and Figure B.2. The REA curve for Cellets and the material specific regression function are depicted in Figure 4.14. $R^2$ being larger than 99 % confirms the collapse of all data points into one material specific curve.

Drying experiments with FCC catalyst are conducted in the pilot-plant-scale VFB dryer. Mechanical vibration is varied in the parameter studies in addition to the above presented parameters. The measured drying kinetics of FCC catalyst are plotted against time in Figure 4.15. The data points are more scattered, compared to Cellets and $\gamma$-$Al_2O_3$ particles. The different measurement methods of the moisture content are the underlying cause. Inline measurement of particle moisture content are performed for $\gamma$-$Al_2O_3$ particles and Cellets (in the *GF3* dryer). However, the inline measurement requires the sample to be stationary in the sensor head for a few seconds. Hence, vibration of the dryer prevents inline measurement. Particle moisture content needs to be measured offline. The time resolution of the inline measurement is much higher, than the offline measurements in the VFB dryer (see chapter 3.4.1), explaining the scattering of the data for FCC catalyst. Nonetheless, Figure 4.15 depicts increasing drying time with increasing bed mass. In all cases, the critical moisture content is approximately 0.05 kg/kg and the equilibrium moisture content is 0.02 kg/kg.

Regarding the influence of vibration on the drying kinetics, no clear trend is evident in the investigated parameter range, implying that the investigated vibration parameters have no significant effect on the drying kinetics of FCC catalyst. Consequently, the investigated vibration parameters do not affect the hydrodynamics or agglomerate size, compared to the non-vibrated case. However, the reduction of $u_{mf}$ by mechanical vibra-

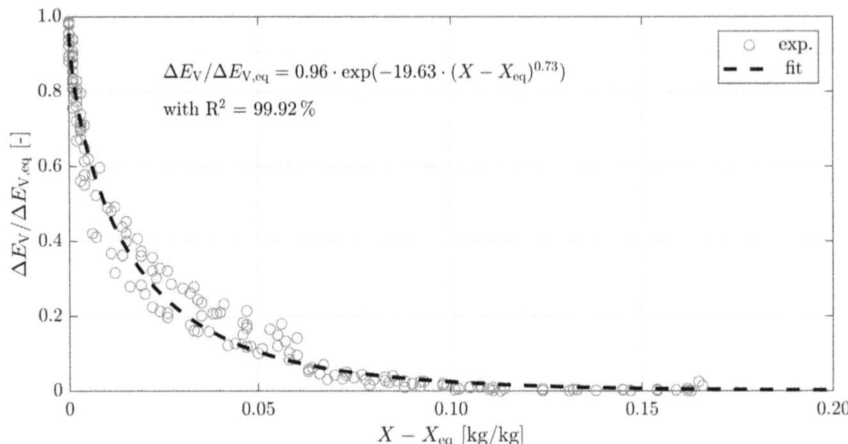


Figure 4.14: Reaction Engineering Approach data (normalized activation energy $\Delta E_V/\Delta E_{V,eq}$) against moisture content $(X - X_{eq})$ for all experiments with Cellets, including material specific regression function.

tion is evident for FCC catalyst as well as Cellets. This has been discussed in chapter 4.1 and was shown for other powders (Mawatari et al., 2015).

The REA curves for all experiments with FCC catalyst are illustrated in Figure 4.16. It shows scattering of data points of the individual experiments in the second drying period. This is to be expected, considering the scattered drying kinetics data, due to the lower time resolution of the offline measurements. Nonetheless, the $R^2$ value larger than 99 % indicates reliable determination of the material specific REA curve of FCC catalyst.

The determined material specific REA curves are used in the flowsheet simulations in the scope of this work.

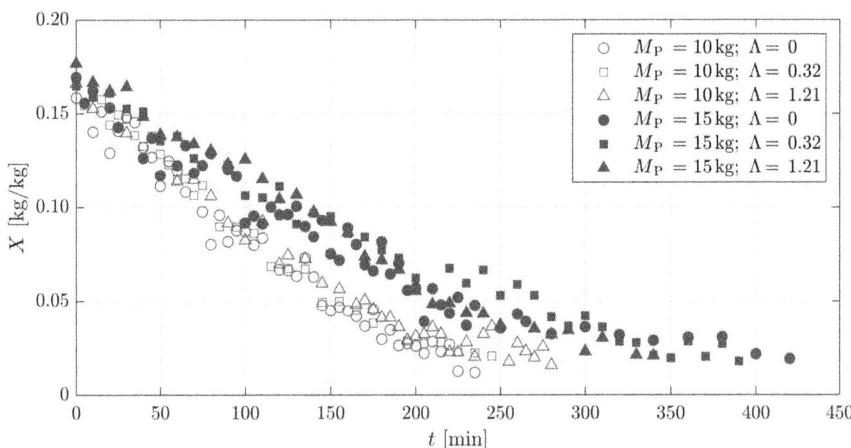


Figure 4.15: Moisture content $X$ of FCC catalyst against drying time ($t$), showing influence of bed mass ($M_P$) and vibration intensity $\Lambda$ on drying kinetics in the VFB dryer.

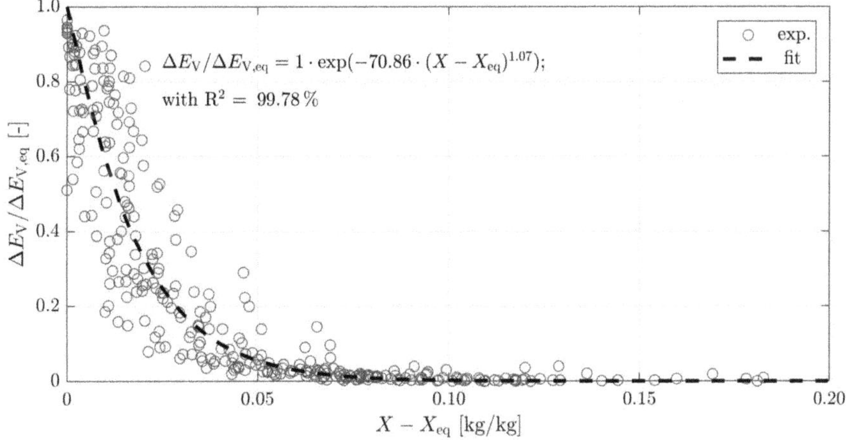


Figure 4.16: Reaction Engineering Approach data (normalized activation energy $\Delta E_V/\Delta E_{V,eq}$) against moisture content ($X - X_{eq}$) for all experiments with FCC catalyst, including material specific regression function.

## 4.4 Model Implementation

The developed model is implemented in the flowsheet simulation framework DYSSOL. An interface for the SUNDIALS package and KINSOL solver is provided by DYSSOL (Lawrence Livermore National Laboratory, 2020, Skorych et al., 2020). The solver utilizes fixed-point iterations with Anderson Acceleration. In order to identify variable updates in the subsequent iteration, intermediate solution steps are executed in combination with computed weights (Anderson, 1965). Integrals are computed with the trapezoidal rule. All relevant material properties are available in DYSSOL's material database, representing the material properties' dependence on process parameters (temperature, pressure and aggregate state) in form of tables. During simulations, material properties are accessed by the model and determined via interpolation.
The schematic flowchart of the model algorithm is illustrated in Figure 4.17. The model consists of four subsequent steps: *Initialization*, *Hydrodynamic Model*, *Thermodynamic Model* and *Post Processing*.
In the *Initialization* step, solver parameters, boundary and initial conditions for both calculation steps are computed. Initial conditions of the hydrodynamic model are initial bubble diameter and total bed height. Enthalpy of gas and particle phase are the initial conditions of the thermodynamic model, which are based on temperature and moisture distribution of the inlet streams.
The *Hydrodynamic Model* calculates the averaged bubble diameter, bubble volume fraction $\varepsilon_{\mathrm{B}}$ and bed height $H_{\mathrm{fb}}$. The underlying correlations are chosen based on the respective Geldart group of the particles and are introduced in detail in chapter 4.4.1. A comprehensive sensitivity analysis of the fluidized bed drying model identified that the hydrodynamics has low impact of changes on the model predictions (see chapter 5.2). Thus, to reduce the simulation time and increase computational robustness, the *Hydrodynamic Model* is calculated only once, and the results are passed on to the *Thermodynamic Model.*
Temperature and moisture content of the phases are calculated by solving mass and energy balances of the different phases and the underlying transfer streams in the *Thermodynamic Model.* Heat and mass balances as well as transfer streams are introduced in chapter 4.4.2 and schematically shown in Figure 4.18.
In the *Post Processing* step, the model results are passed to the outlet streams and the simulation data is processed for further analysis within DYSSOL or exported for external use.

### 4.4.1 Hydrodynamic Model

Fluidized bed hydrodynamics depend on the bubble characteristics, which vary for different Geldart groups. Analogous to previous studies (Alaathar, 2017, Burgschweiger, 2000, Groenewold and Tsotsas, 1999), the hydrodynamics model from Hilligardt and Werther (1986) is used for non-cohesive particles of Geldart groups A, B and D. The impact

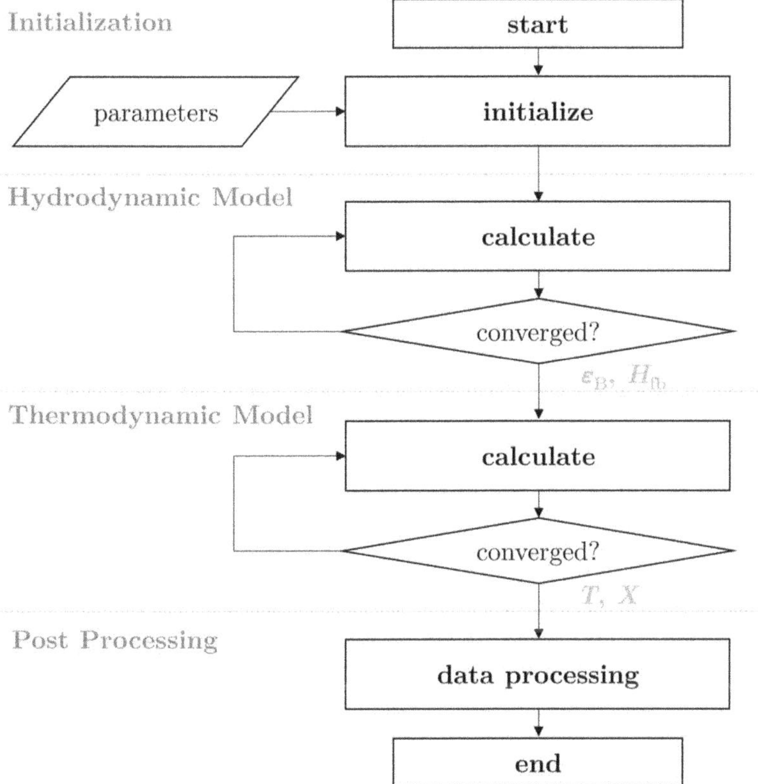


Figure 4.17: Schematic flowchart of the algorithm of the vibrated fluidized bed drying, implemented in DYSSOL.

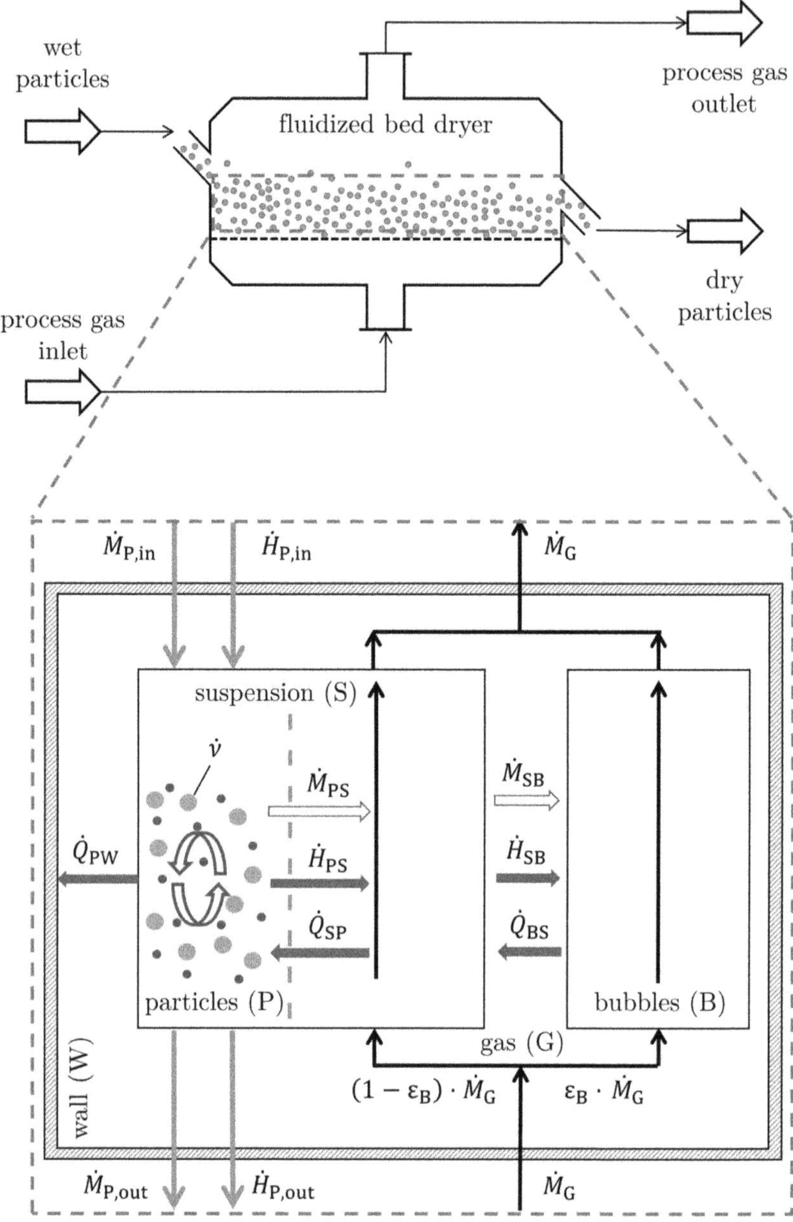


Figure 4.18: Schematic flowchart of the model for vibrated fluidized bed drying, including three distinct phases: ideally mixed particle phase (P), plug flow bubbles (B) and suspension gas (S), as well as the respective heat($\dot{Q}$), mass ($\dot{M}$) and enthalpy ($\dot{H}$) transfer streams.

of vibration is considered by including the vibration intensity and measured values of minimum fluidization velocity and bed porosity at minimum fluidization, as introduced in chapter 4.1. Hydrodynamics of cohesive Geldart A and C particles are represented by the correlations, developed in chapter 4.1. In all cases, the bubble volume fraction is calculated using equations 4.1 through 4.3:

$$\varepsilon_{\mathrm{B}} = \frac{\dot{V}_{\mathrm{B}}}{u_{\mathrm{B}}}, \tag{4.1}$$

$$\dot{V}_{\mathrm{B}} = \psi \cdot (u - u_{\mathrm{mf}}) \cdot \frac{1}{1+\Lambda}, \tag{4.2}$$

$$u_{\mathrm{B}} = \dot{V}_{\mathrm{B}} + 0.71 \cdot \Theta \sqrt{g \cdot d_{\mathrm{B}}}\,. \tag{4.3}$$

The hydrodynamic parameters $\psi$ and $\Theta$ depend on the Geldart group. Furthermore, $\psi$ depends on the ratio of vertical position in the bed and bed diameter. Also, $\Theta$ depends on the diameter of the fluidized bed. Respective values are listed in Table 4.3. The bubble diameter $d_{\mathrm{B}}$ changes with height above the distributor $z$ and differs between Geldart groups.

The bubble diameter in non-cohesive particles is described by the following correlation from Hilligardt and Werther (1987):

$$\frac{\partial(d_{\mathrm{B}})}{\partial z} = \left(\frac{2 \cdot \varepsilon_{\mathrm{B}}}{9\pi}\right)^{1/3} \left(1 - \kappa \cdot \left(\frac{6\varepsilon_{\mathrm{B}}}{\pi}\right)^{1/3}\right)^{-1} - \frac{d_{\mathrm{B}}}{3 \cdot \lambda \cdot u_{\mathrm{B}}}\,. \tag{4.8}$$

Here, the first part describes bubble growth due to coalescence. The second part accounts for splitting of bubbles, using the average bubble life time $\lambda$:

$$\lambda = 280 \cdot \frac{u_{\mathrm{mf}}}{g}\,. \tag{4.9}$$

The factor $\kappa$ accounts for dynamics bubble coalescence (Alaathar, 2017):

$$\kappa = \begin{cases} 1 - k^3 & \text{for } k \leq 1 \quad \text{'slow bubbles'} \\ 0 & \text{for } k > 1 \quad \text{'fast bubbles'}\,. \end{cases} \tag{4.10}$$

It depends on $k$, expressing the bubble velocity relative to the velocity of the suspension gas:

$$k = \frac{u_{\mathrm{B}} \cdot \varepsilon_{\mathrm{S}}}{u_{\mathrm{S}}}\,. \tag{4.11}$$

The porosity of the suspension phase $\varepsilon_{\mathrm{S}}$ is calculated with equation 4.6. According to Alaathar (2017), $u_{\mathrm{S}}$ in non-cohesive particle beds is influenced by the type of distributor:

$$u_{\mathrm{S}} = \begin{cases} 1/4 \cdot (u - u_{\mathrm{mf}}) + u_{\mathrm{mf}} & \text{for industrial distributors} \\ 1/3 \cdot (u - u_{\mathrm{mf}}) + u_{\mathrm{mf}} & \text{for porous plate distributors.} \end{cases} \tag{4.12}$$

In analogy to Alaathar (2017), the differential equation 4.8 for the bubble diameter is solved in dependence of height above the distributor $z$ with the an initial bubble diameter $d_{\mathrm{B},0}$ at the distributor ($z = 0$):

$$d_{\mathrm{B},0} = 1.3 \left( \frac{\dot{V}_{\mathrm{or}}^2}{g} \right)^{0.2} . \tag{4.13}$$

Here, $\dot{V}_{\mathrm{or}}$ denotes the volumetric flow rate of gas through one orifice of the distributor, depending on overall gas velocity $u$, cross sectional area of the bed $A$ and the number of orifices in the distributor $N_{\mathrm{or}}$:

$$\dot{V}_{\mathrm{or}} = \frac{A \cdot u}{N_{\mathrm{or}}} . \tag{4.14}$$

In case porous plate distributor are used, $d_{\mathrm{B},0}$ is assumed to be 0.001 m.

**Geldart group D:**

Geldart group D particles are characterized by high $u_{\mathrm{mf}}$, resulting in long bubble life time after equation 4.9. According to Hilligardt and Werther (1987), it can be assumed that the splitting term in equation 4.8 becomes negligibly small. Hence, the bubble diameter in fluidized beds of Geldart D particles is calculated by:

$$\frac{\partial(d_{\mathrm{B}})}{\partial z} = \left( \frac{2 \cdot \varepsilon_{\mathrm{B}}}{9\pi} \right)^{1/3} \left( 1 - \kappa \cdot \left( \frac{6\varepsilon_{\mathrm{B}}}{\pi} \right)^{1/3} \right)^{-1} . \tag{4.15}$$

**Geldart group A and B:**

In principle, bubbles in fluidized beds of Geldart groups A and B particles fall under the category 'fast bubbles', meaning they rise faster than the suspension gas (Hilligardt and Werther, 1987). Consequently, $\kappa$ becomes 0 and equation 4.8 is reduced to:

$$\frac{\partial(d_{\mathrm{B}})}{\partial z} = \left( \frac{2 \cdot \varepsilon_{\mathrm{B}}}{9\pi} \right)^{1/3} - \frac{d_{\mathrm{B}}}{3 \cdot \lambda \cdot u_{\mathrm{B}}} . \tag{4.16}$$

**Geldart group C and cohesive group A:**

For Geldart group C and cohesive particles of group A, the bubble diameter is calculated with equation 4.4 from Zou et al. (2011), as introduced in chapter 4.1:

$$d_{\mathrm{B}} = 0.21 \frac{(u - u_{\mathrm{mb}})^{0.49} \left( z + 4\sqrt{A_0} \right)^{0.48}}{g^{0.2}} . \tag{4.4}$$

The expansion of the bed and the resulting bed height are also calculated in dependence of the respective particle Geldart classifications. For cohesive group A particles and group C particles the newly developed correlations are used (equations 4.5, 4.6 and 4.7).

Table 4.3: Dimensionless hydrodynamic parameters for calculation of bubble characteristics according to Alaathar (2017), Burgschweiger (2000), Hilligardt and Werther (1986), for Geldart groups A, B and D. With $z$ being the vertical distance from the distributor and $d_{\mathrm{fb}}$ depicting the diameter of the bed. In case of rectangular fluidized beds, $d_{\mathrm{fb}}$ refers to the shortest distance between walls.

| Geldart group | $\psi$ [-] | valid for | $\Theta$ [-] | valid for |
|---|---|---|---|---|
| A | 0.8 | $\frac{z}{d_{\mathrm{fb}}} \leq 1$ | 1.18 | $d_{\mathrm{fb}} \leq 0.05\,\mathrm{m}$ |
| | | | $3.2 \cdot d_{\mathrm{fb}}^{1/3}$ | $0.05 \leq d_{\mathrm{fb}} \leq 1\,\mathrm{m}$ |
| | | | 3.2 | $d_{\mathrm{fb}} > 1\,\mathrm{m}$ |
| B | 0.67 | $\frac{z}{d_{\mathrm{fb}}} \leq 1.7$ | 0.63 | $d_{\mathrm{fb}} \leq 0.1\,\mathrm{m}$ |
| | $0.51 \left(\frac{z}{d_{\mathrm{fb}}}\right)^{1/2}$ | $1.7 \leq \frac{z}{d_{\mathrm{fb}}} \leq 4$ | $2 \cdot d_{\mathrm{fb}}^{1/2}$ | $0.1 \leq d_{\mathrm{fb}} \leq 1\,\mathrm{m}$ |
| | 1 | $\frac{z}{d_{\mathrm{fb}}} > 4$ | 2 | $d_{\mathrm{fb}} > 1\,\mathrm{m}$ |
| C | 1 | n.s.* | 1.18 | $d_{\mathrm{fb}} \leq 0.05\,\mathrm{m}$ |
| | | | $3.2 \cdot d_{\mathrm{fb}}^{1/3}$ | $0.05 \leq d_{\mathrm{fb}} \leq 1\,\mathrm{m}$ |
| | | | 3.2 | $d_{\mathrm{fb}} > 1\,\mathrm{m}$ |
| D | 0.26 | $\frac{z}{d_{\mathrm{fb}}} \leq 0.5$ | 0.87 | n.s.* |
| | $0.35 \left(\frac{z}{d_{\mathrm{fb}}}\right)^{1/2}$ | $0.55 \leq \frac{z}{d_{\mathrm{fb}}} \leq 8$ | | |
| | 1 | $\frac{z}{d_{\mathrm{fb}}} \leq 8$ | | |

*not specified

For non-cohesive particles of Geldart groups A, B and D, the bed expansion is calculated using the correlation introduced by Richardson and Zaki (1954) and modified to account for the impact of vibration via vibration intensity:

$$\varepsilon = \varepsilon_{\mathrm{mf}} \cdot \left(\frac{u_{\mathrm{S}}}{u_{\mathrm{mf}}}\right)^{1/(4.65 \cdot (1+\Lambda))}, \tag{4.17}$$

using equation 4.7 for computation of $u_{\mathrm{S}}$. The bed height of the operational fluidized bed results from the bed expansion ratio and the bed height of the fixed bed $H_{\mathrm{fix}}$:

$$H_{\mathrm{fb}} = \frac{1 - \varepsilon_0}{1 - \varepsilon} \cdot H_{\mathrm{fix}} . \tag{4.18}$$

### 4.4.2 Thermodynamic Model

Thermodynamic properties of the three distinct phases, as defined in Figure 4.18, are calculated in the thermodynamic model, based on the bubble volume fraction $\varepsilon_{\mathrm{B}}$ and the bed height $H_{\mathrm{fb}}$ from the hydrodynamics model. Since steady state operations are considered in this work, the mass flow rate of dry fluidization gas is independent of the position in the bed and the particle mass flow on dry basis is independent of residence time in the dryer. Hence, the respective mass balances only cover water, transferred during the drying process.

### Balance Equations

As introduced in chapter 4.4, the gas moves as plug flow. Thus, the gas phases are calculated in dependence of the normalized height coordinate $\zeta = z/H_{\text{fb}}$. Balance equations of suspension and bubble phase are:

$$(1-\varepsilon_{\text{B}}) \cdot \dot{M}_{\text{G}} \cdot \frac{dY_{\text{S}}}{d\zeta} = \frac{\partial}{\partial\zeta}\left(\dot{M}_{\text{PS}} - \dot{M}_{\text{SB}}\right), \tag{4.19}$$

$$(1-\varepsilon_{\text{B}}) \cdot \dot{M}_{\text{G}} \cdot \frac{dh_{\text{S}}}{d\zeta} = \frac{\partial}{\partial\zeta}\left(\dot{H}_{\text{PS}} + \dot{Q}_{\text{PS}} - \dot{H}_{\text{SB}} - \dot{Q}_{\text{SB}}\right), \tag{4.20}$$

$$\varepsilon_{\text{B}} \cdot \dot{M}_{\text{G}} \cdot \frac{dY_{\text{B}}}{d\zeta} = \frac{\partial}{\partial\zeta}\dot{M}_{\text{SB}}, \tag{4.21}$$

$$\varepsilon_{\text{B}} \cdot \dot{M}_{\text{G}} \cdot \frac{dh_{\text{B}}}{d\zeta} = \frac{\partial}{\partial\zeta}\left(\dot{H}_{\text{SB}} + \dot{Q}_{\text{SB}}\right). \tag{4.22}$$

Properties of the particle phase depend on residence time $\tau$ as well as on distributed parameters. Distributed parameters are discretized using the method of classes with respect to particle moisture content (subscript $f$) and particle size (subscript $d$). Residence time and normalized height are discretized via finite volume method. The balance equations of the particle phase are described by:

$$M_{\text{P},d,f} \cdot n(\tau) \cdot \frac{dX_{d,f}}{d\tau} = -\int_0^1 \frac{\partial^2}{\partial\tau\partial\zeta}\dot{M}_{\text{PS},d,f}\, d\zeta, \tag{4.23}$$

$$M_{\text{P},d,f} \cdot n(\tau) \cdot \frac{dh_{d,f}}{d\tau} = -\int_0^1 \frac{\partial^2}{\partial\tau\partial\zeta}\left(\dot{H}_{\text{PS},d,f} + \dot{Q}_{\text{PS},d,f}\right) d\zeta + \frac{\partial}{\partial\tau}\dot{Q}_{\text{PW},d,f}. \tag{4.24}$$

The residence time dependent number distribution $n$ is defined with the tank in series (TIS) model, as introduced in chapter 2.1.4.

$$n(\tau) = \frac{1}{\bar{\tau}}\left(\frac{\tau}{\bar{\tau}}\right)^{K-1} \frac{K^K}{(1-K)!}\exp\left(-\tau K/\bar{\tau}\right) \tag{4.25}$$

Here, $\bar{\tau}$ denotes the hydrodynamic residence time, under steady state conditions:

$$\bar{\tau} = \frac{M_{\text{P}}}{\dot{M}_{\text{P}}}, \text{ with } \dot{M}_{\text{P,out}} = \dot{M}_{\text{P,in}} = \dot{M}_{\text{P}}\,. \tag{4.26}$$

Application of the gamma function $\Gamma(K)$ allows for simulations of continuous fluidized bed dryers, in which the particle residence time distribution is represented by non-integer numbers of tanks $K$. The gamma function is defined as the improper integral, valid for complex number with positive real parts (Davis, 1959):

$$(1-K)! = \Gamma(K) = \int_0^\infty x^{K-1}e^{-x}dx\,. \tag{4.27}$$

Equation 4.25 asymptotically approaches zero. Consequently, the discretization requires the definition of a maximum value for the residence time ($\tau_{max}$). By default, this value is calculated, considering 99 % of the particles, i.e.:

$$\tau_{max} = -\log(1-0.99)\,\frac{M_{\mathrm{P}}}{\dot{M}_{\mathrm{P}}}\,. \tag{4.28}$$

**Transfer Equations:**

All transfer streams between the phases are established in analogy to Alaathar (2017). Mass transfer streams between the phases are defined as follows:

$$\frac{\partial}{\partial\zeta}\dot{M}_{\mathrm{SB}} = \rho_{\mathrm{G}}\cdot\beta_{\mathrm{SB}}\cdot A_{\mathrm{SB}}\left(Y_{\mathrm{S}}(\zeta)-Y_{\mathrm{B}}(\zeta)\right), \tag{4.29}$$

$$\frac{\partial^2}{\partial\zeta\partial\tau}\dot{M}_{\mathrm{PS},d,f} = \dot{\nu}\cdot\rho_{\mathrm{G}}\cdot\beta_{\mathrm{PS},d,f}\cdot n(\tau)\cdot A_{\mathrm{PS},d,f}\left(Y_{\mathrm{eq}}(X,T_{\mathrm{P}})-Y_{\mathrm{S}}(\zeta)\right), \tag{4.30}$$

where:

$$\frac{\partial}{\partial\zeta}\dot{M}_{\mathrm{PS}} = \int_0^{\tau_{\max}}\sum_{d=0}^{N_d-1}\sum_{f=0}^{N_f-1}\frac{\partial^2}{\partial\zeta\partial\tau}\dot{M}_{\mathrm{PS},d,f}\,d\tau\,. \tag{4.31}$$

In equation 4.30, the material specific drying curve $\dot{\nu}$ is considered. It may be derived via the inverted Reaction Engineering Approach (REA) (see chapter 4.3.1) or the traditional Normalized Characteristic Drying Curve (NCDC), alike. Furthermore, $Y_{\mathrm{eq}}(X, T_{\mathrm{P}})$ represents the equilibrium moisture content of air in the vicinity of the particles and is calculated with equation 2.34, using the relative humidity of air in equilibrium, based on the measured sorption isotherms (see Figure 3.5).
Enthalpy transfer streams are described by:

$$\frac{\partial}{\partial\zeta}\dot{H}_{\mathrm{SB}} = c_{p,\mathrm{H_2O,v}}\cdot(T_{\mathrm{SB}}-T_0)\Delta h_{v,\mathrm{H_2O},0}\frac{\partial}{\partial\zeta}\dot{M}_{\mathrm{SB}}, \tag{4.32}$$

$$\frac{\partial^2}{\partial\zeta\partial\tau}\dot{H}_{\mathrm{PS},d,f} = \left(c_{p,\mathrm{H_2O,v}}\cdot(T_{\mathrm{PS},d,f}-T_0)+\Delta h_{v,\mathrm{H_2O},0}\right)\frac{\partial^2}{\partial\zeta\partial\tau}\dot{M}_{\mathrm{PS},d,f}, \tag{4.33}$$

with:

$$\frac{\partial}{\partial\zeta}\dot{H}_{\mathrm{PS}} = \int_0^{\tau_{\max}}\sum_{d=0}^{N_d-1}\sum_{f=0}^{N_f-1}\frac{\partial^2}{\partial\zeta\partial\tau}\dot{H}_{\mathrm{PS},d,f}\,d\tau \tag{4.34}$$

and $T_0 = 273.15\,\mathrm{K}$ denoting the reference temperature and $\Delta h_{\mathrm{v,H_2O},0}$ the enthalpy of evaporation of water at $T_0$ and the governing pressure. The pressure is set to 1.013 25 bar throughout this work, but may easily be adjusted within the DYSSOL GUI for other cases.

Heat transfer streams are denoted as:

$$\frac{\partial}{\partial \tau}\dot{Q}_{\mathrm{PW},d,f} = \alpha_{\mathrm{PW},d,f} \cdot n(\tau) \cdot A_{\mathrm{PW}} \cdot (T_{\mathrm{W}} - T_{\mathrm{P},d,f}(\tau)), \tag{4.35}$$

$$\frac{\partial}{\partial \zeta}\dot{Q}_{\mathrm{SB}} = \alpha_{\mathrm{SB}} \cdot A_{\mathrm{SB}} \left(T_{\mathrm{S}}(\zeta) - T_{\mathrm{B}}(\zeta)\right), \tag{4.36}$$

$$\frac{\partial^2}{\partial \zeta \partial \tau}\dot{Q}_{\mathrm{PS},d,f} = \alpha_{\mathrm{PS},d,f} \cdot A_{\mathrm{PS},d,f} \left(T_{\mathrm{P},d,f}(\tau) - T_{\mathrm{S}}(\zeta)\right), \tag{4.37}$$

with:

$$\frac{\partial}{\partial \zeta}\dot{Q}_{\mathrm{PS}} = \int_0^{\tau_{\max}} \sum_{d=0}^{N_d-1} \sum_{f=0}^{N_f-1} \frac{\partial^2}{\partial \zeta \partial \tau}\dot{Q}_{\mathrm{PS},d,f}\, d\tau. \tag{4.38}$$

In Equation 4.35, $A_{\mathrm{PW}}$ is the surface area of the dryer wall that is in contact with the particles. Mass transfer between particles and walls can be neglected due to short contact times (Alaathar et al., 2013, Martin, 2010).
Heat and mass transfer coefficients are calculated after established correlations by Gnielinski (1980), Martin (2010), Tsotsas (2010), which are listed in Appendix C.

### Initial Conditions

The following initial conditions are defined:

$$X_{\mathrm{P}}\,(\tau = 0) = X_{\mathrm{P,in}}\,, \tag{4.39}$$

$$T_{\mathrm{P}}\,(\tau = 0) = T_{\mathrm{P,in}}\,, \tag{4.40}$$

$$Y_{\mathrm{B}}\,(\zeta = 0) = Y_{\mathrm{S}}\,(\zeta = 0) = Y_{\mathrm{in}}\,, \tag{4.41}$$

$$T_{\mathrm{B}}\,(\zeta = 0) = T_{\mathrm{S}}\,(\zeta = 0) = T_{\mathrm{G,in}}\,. \tag{4.42}$$

# 5

# Simulation Results

Comprehensive validation of the new model is performed by comparison of experimental results with model predictions. Model validation for different fluidized bed dryers and materials is discussed in the following chapters. Sensitivity analysis of the model is topic of chapter 5.2. Parts of these results have been previously published in Lehmann et al. (2021).
All simulations for validation and sensitivity analysis were conducted on a regular PC (Windows 10 Pro x64, 16GB RAM, Intel Core i7-6700). The duration of the simulations varied between one and ten minutes.

## 5.1 Validation

The validation of the model is divided into different aspects. The first validation aspect covers Geldart D and B particles, investigated in the lab-scale dryer in chapters 5.1.1 and 5.1.2. In analogy to Alaathar (2017), the residence time distribution (RTD) is assumed to match ideal CSTR behavior in the lab-scale *CF3* dryer. Investigated cases and respective process conditions are listed in Table 5.1.
In a second step, the model predictions are tested for a different dryer size and geometry, using the VFB dryer. Effects of RTD, deviating from ideal CSTR characteristics, are investigated in chapters 5.1.3 and 5.2.3. Mechanical vibration is studied as an additional process parameter in chapter 5.1.3.
The third part of the validation treats Geldart A particles, investigated in the VFB dryer (chapter 5.1.4). An overview of all experimental and simulation results, used for validation, are summarized in Table D.1 and Table D.2. Mass weighted averages of predicted particle temperature and moisture content are derived for comparison with measurements from the bulk of the samples.

One aspect of the model that requires special attention is the estimation of heat transfer from dryer to environment. Conventional calculation of heat transfer from dryer to environment requires accurate determination of the heat transfer resistances of the dryer walls and possible thermal insulation. Said transfer resistances are highly specific to the investigated facility and thus, are likely to differ significantly from dryer to dryer. Depending on the degree of insulation at different parts (e.g. windbox, drying chamber, free board and pipes), transfer resistances may even vary for different areas of a single dryer.

A different approach is implemented in the proposed model, allowing for a more general application of the model without knowledge of details about the dryer's insulation (see equation 4.35). In this model, heat transfer from the drying chamber to the environment is estimated by considering heat transfer from the particles to the walls of the drying chamber. Hence, the temperature of the inside of the dryer wall is required. It is considered in the model as an input parameter. However, the dryers used in the experiments of this work are not equipped for the measurement of wall temperatures. Thus, the wall temperature would be based on assumptions. Investigated drying temperatures vary between 40 and 80 °C. Based on the relatively low temperature difference between drying air and ambient, the influence of heat transfer from the dryer to environment may be expected to be minor, compared to other involved heat transfer mechanisms. Therefore, heat loss from the dryer to the environment in form of heat transfer between particles and dryer wall is not considered in the simulations used for the general model validation. This assumption and its range of validity is discussed within the following chapters.

### 5.1.1 Validation for Geldart D Particles

Model predictions are compared to experiments, conducted with $\gamma$-$Al_2O_3$ particles in the *GF3* lab-scale dryer. Height of the outflow hole and drying gas temperature are varied. Operation of the *GF3* dryer with ambient air entails a dependency of the results on inlet moisture content of the drying air $Y_{\text{in}}$. Additional parameter studies are conducted with varied inlet moisture content of particles $X_{\text{P,in}}$ and particle mass flow $\dot{M}_{\text{P,dry}}$.

Validation of the hydrodynamics model is performed by comparing calculated and measured bed heights. These are plotted against drying temperature and shown in Figure 5.1. Bed height predictions lie within the uncertainty of experiments, depicted by error bars. Therefore, the accuracy of the hydrodynamic model predictions is confirmed. The results also show that measured and calculated bed heights are lower than the outlet height. This is based on the bubbling character of the fluidized bed. While bubbles erupt at the bed surface, particles are thrown upwards. A fraction of these particles leave the dryer through the outlet hole, leading to bed heights below the outflow hole.

Overall model performance is validated based on the accuracy of predicted particle temperature $T_{\text{P}}$ and moisture content $X_{\text{P}}$. Variations in ambient air conditions, specifically inlet moisture content of air $Y_{\text{in}}$, necessitates three dimensional visualization of comparisons of experimental and simulation data. Validation data are plotted against drying temperature and inlet moisture content of drying air. Gray planes depict the simulation data. Red and blue planes illustrate the lower and upper limit of the experimental margin of error of particle temperature and moisture content, respectively.

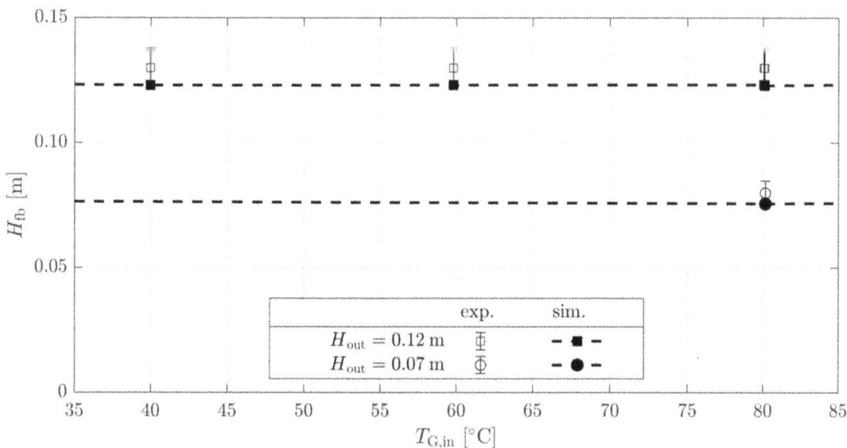


Figure 5.1: Comparison of predicted bed height ($H_{fb}$) and measurements in *GF3* lab-scale dryer for $\gamma$-$Al_2O_3$ particles, plotted against gas temperature ($T_{G,in}$) for different outflow hole heights ($H_{out}$).

Figure 5.2 shows measured and predicted particle temperature under different process conditions. It is evident that increasing drying temperature strongly influences the particle temperature. Comparison of cases GF3_A_04 and GF3_A_05 demonstrates that particle temperature is not affected by the inlet moisture content of drying air, given similar ratios of particle mass flow and particle inlet moisture content. Both observations are expected and physically founded. Case GF3_A_02 differs from the other depicted cases, as the particles are still in the first drying period. Therefore, particle temperature equals the wet bulb temperature, as measured and predicted by the model alike.

Overall, measured and predicted particle temperature are in good agreement (see Figure 5.2). Relative deviations never exceed 4 %. Increasing relative deviations are observed between experiments and simulations with increasing drying temperature. At drying temperatures of 80 °C the predictions lie outside the experimental uncertainty and the model slightly over-predicts particle temperature. Likely, this is based on the fact that heat transfer between dryer and environment is neglected in these simulations.

Including heat transfer between particles and dryer wall in the simulations and fitting the assumed wall temperature to optimize model predictions may result in more accurate estimations. However, application of such case specific fitting parameters would strongly limit models applicability to other cases, such as other conditions, different dryer configurations or varying materials. This topic is addressed thoroughly in chapter 5.2.4. Considering the good agreement between measured and predicted particle temperature in light of the goal of a universal model, validation is continued without consideration

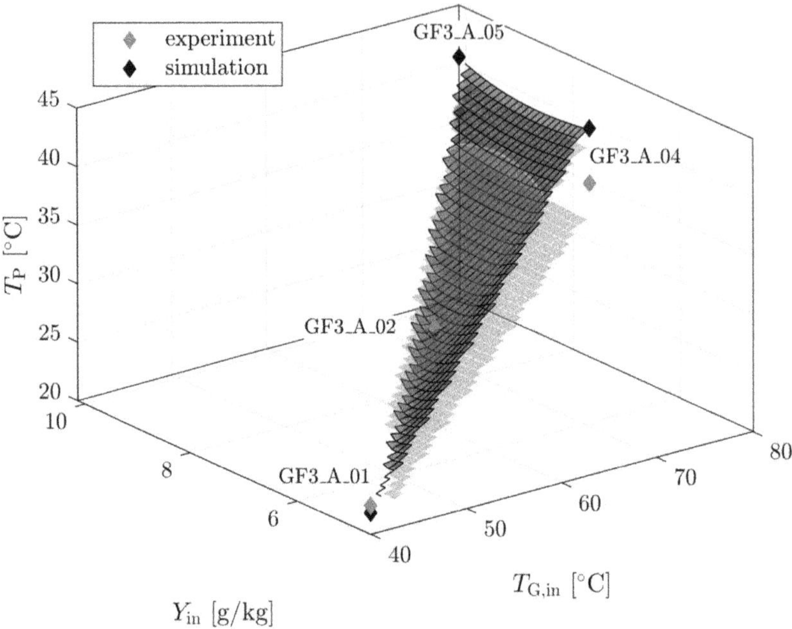


Figure 5.2: Comparison of measured and predicted particle temperature ($T_P$), plotted against drying temperature ($T_{G,in}$) and inlet moisture content of air ($Y_{in}$) in *GF3* lab-scale dryer for $\gamma$-$Al_2O_3$ particles. Dark gray area designates model predictions; light red areas indicate the upper and lower boundary of measurement errors. Detailed process conditions of represented data points are given in Table 5.1.

of heat transfer from dryer to environment.

Particle moisture content from experiments and simulations are illustrated in Figure 5.3. The results display increased particle moisture content at the outlet for increased inlet particle mass flow, as expected. In case GF3_A_02, the combination of higher particle mass flow and higher particle inlet moisture content outweighs the effect of reduced particle moisture content with increased drying temperature, which is evident in all other depicted cases. Additionally, experiments and simulations exhibit the expected reduction of particle moisture content for lower moisture content of inlet air. All model predictions of $X_P$ lie within the experimental margin of error, attributing high accuracy of the model for the investigated range of drying temperatures, moisture contents of inlet air, bed heights as well as particle mass flow rates. Relative deviation between experiment and model is approximately 15 % for case GF3_A_01 (lowest drying temperature, inlet moisture content of air and particles). In all other cases, relative deviations are lower than 6 %.

The discussed validation results certify accurate representation of physical effects on the particle properties in the investigated range of process parameters by the model for

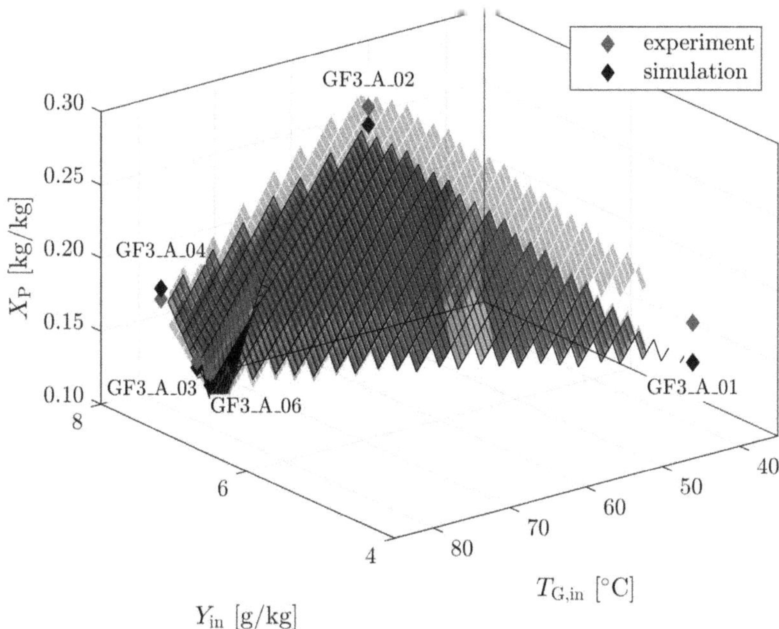


Figure 5.3: Comparison of measured and predicted particle moisture content ($X_P$), plotted against drying temperature ($T_{G,in}$) and inlet moisture content of air ($Y_{in}$) in *GF3* lab-scale dryer for $\gamma$-$Al_2O_3$ particles. Dark gray area designates model predictions; light blue areas indicate the upper and lower boundary of measurement errors. Detailed process conditions of represented data points are given in Table 5.1.

Geldart D particles. Predicted properties of the drying gas at the outlet of the dryer are compared to experimental data in Figure 5.4. Gas outlet temperature is predicted very similar to experimental data for lower drying temperatures. For elevated drying temperatures however, deviations between calculated and measurements increase. In analogy to particle temperatures, increasing over-prediction by the model is observed and may be explained by potential increased influence of heat transfer between dryer and environment.

The same trend is evident for calculated and measured moisture content of outlet air. However, the deviations are significantly larger. Throughout, simulated gas moisture content is predicted higher than measured in experiments. This discrepancy may be explained by the location of the humidity sensor. In contrast to the bed temperature, measured directly above the bed, the humidity sensor is located in the off gas pipe of the *GF3* dryer (see Figure 3.20). Hence, the moist air passes the expansion zone, the internal filters and a section of the exhaust gas periphery, before its humidity is measured. Along this way from the bed, the air passes un-insulated steel surfaces and piping as well as the filters. Specifically for high moisture content, during the first drying period, conden-

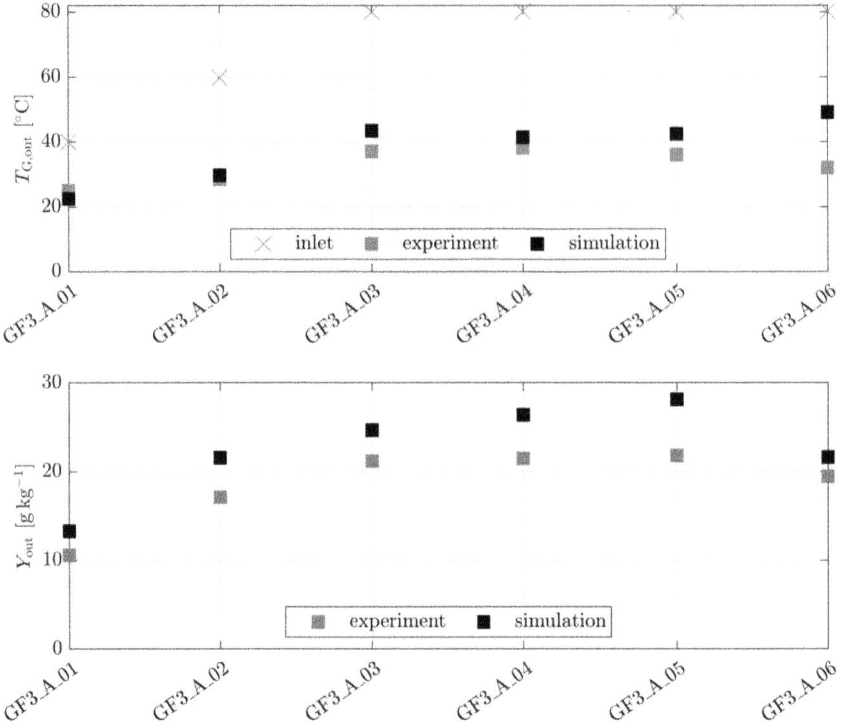


Figure 5.4: Comparison of measured and predicted air temperature ($T_{G,out}$) (top) and moisture content ($Y_{out}$) (bottom) for all validation cases in the *GF3* lab-scale dryer with $\gamma$-$Al_2O_3$ particles; also showing inlet gas temperature. Detailed process conditions of represented data points are given in Table 5.1.

sation of water on said colder surfaces cannot be ruled out and might explain reduced measured moisture content. Additionally, inleaked air observed by Alaathar (2017) in the same setup and discussed in detail in chapter 4.3.1 partially explains the observed deviation between model predictions and measurements. Therefore, the measurement data regarding outlet air moisture content is likely not accurate enough in the *GF3* dryer, to allow for reliable validation. This hypothesis is supported by calculation of water mass balance around the dryer volume within the model. Therein, the mass difference between water entering and leaving the dryer never exceeds 1 %. Consequently, model predictions with respect to gas properties at the outlet are still considered accurate.

### 5.1.2 Validation for Geldart B Particles

Cellets are used as material of Geldart group B. Validation experiments are conducted in the *GF3* lab-scale dryer. Validation results are presented in analogy to the previos chapter. Measured and simulated bed heights are compared for varying superficial gas

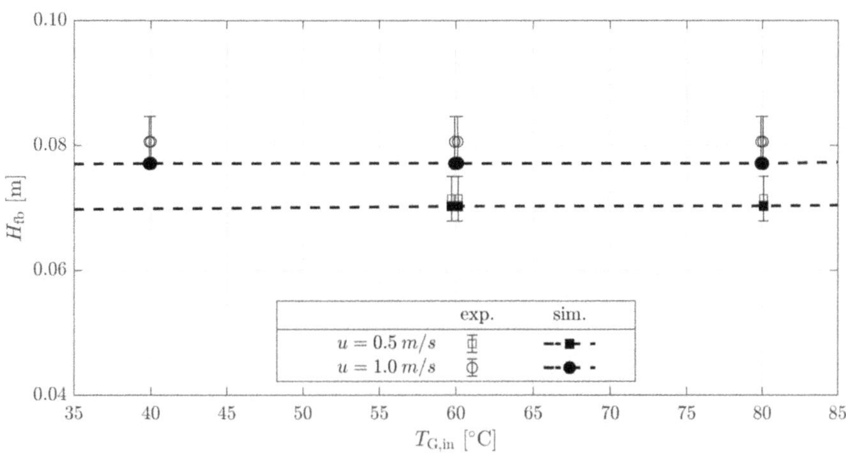


Figure 5.5: Comparison of predicted bed height ($H_{fb}$) and measurements in *GF3* lab-scale dryer for Cellets, plotted against gas temperature ($T_{G,in}$) at different gas velocities ($u$).

velocity $u$ in Figure 5.5. Higher bed expansion is measured and predicted by the model at higher gas velocities, as expected. The height of the outflow hole is constant at 12 cm. Both, observed and calculated bed heights are lower than the height of the outlet hole, for the reasons discussed in chapter 5.1.1. However, the deviation between bed height and height of the outflow hole is larger with Geldart B particles. This is based on more dominant bubble characteristics, namely faster bubble velocity and higher bubble volume fraction in fluidized beds of Geldart B particles. Since predicted bed heights lie within the measurement errors in all investigated cases, this effect is represented successfully by the model, confirming high accuracy of the hydrodynamics model for Geldart B particles.

Comparison of predicted and measured particle temperature under different process conditions is illustrated in Figure 5.6. Particle temperature is plotted against drying temperature and inlet moisture content of drying air. The data depict that increased drying temperature leads to higher particle temperature. In contrast, inlet moisture content of drying air shows negligible impact on the particle temperature in the investigated range and parameter combinations. High accuracy of model predictions is accredited by relative deviations between model and experiments being well below 3 %.
Similar accuracy of simulated particle moisture content is achieved for Cellets in a wide range of drying temperatures and inlet moisture contents of drying air. Figure 5.7 displays the comparison of measured and calculated particle moisture content. All model predictions lie well within the experimental margin of error. Furthermore, expected trends are reflected by model and experiments alike. This being the expected reduction

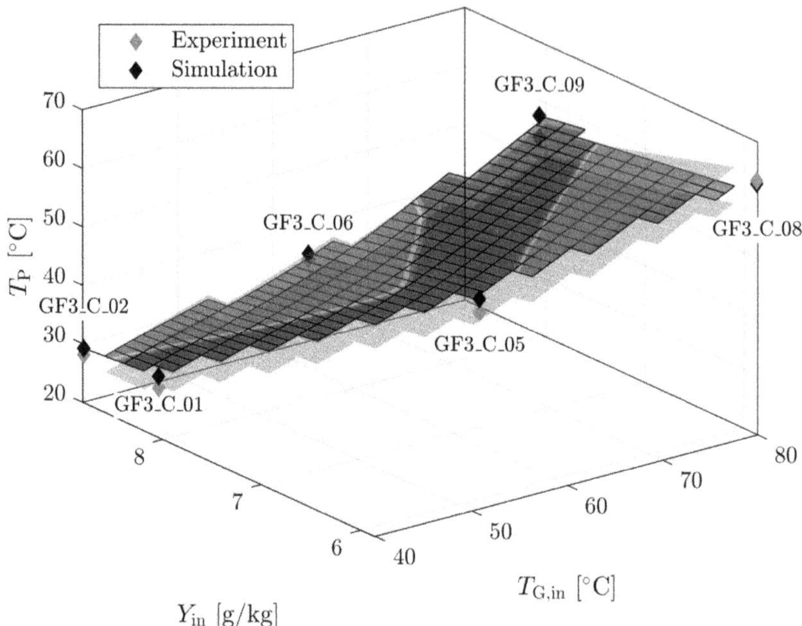


Figure 5.6: Comparison of measured and predicted particle temperature ($T_P$), plotted against drying temperature ($T_{G,in}$) and inlet moisture content of air ($Y_{in}$) in *GF3* lab-scale dryer for Cellets. Dark gray area designates model predictions; light red areas indicate the upper and lower boundary of measurement errors. Detailed process conditions of represented data points are given in Table 5.1.

of particle moisture content at lower air inlet moisture content and higher temperature of drying air.

Thereby, accurate predictions of the proposed model are confirmed with respect to particle properties of Geldart B particles in the examined range of process conditions.

Moisture content and temperature of air at the dryer outlet are compared to measurements in Figure 5.8. It is evident that gas moisture content is predicted accurately. The temperature of outlet gas however, is over-predicted by the model. Significant deviations are observed for cases with higher drying temperature (80 °C). As previously observed and explained for $\gamma$-$Al_2O_3$ particles, the deviations between measured and predicted temperatures at high drying temperatures may be explained by excluding heat transfer between dryer and environment in these validation cases. This topic is investigated thoroughly in chapter 5.2.4.

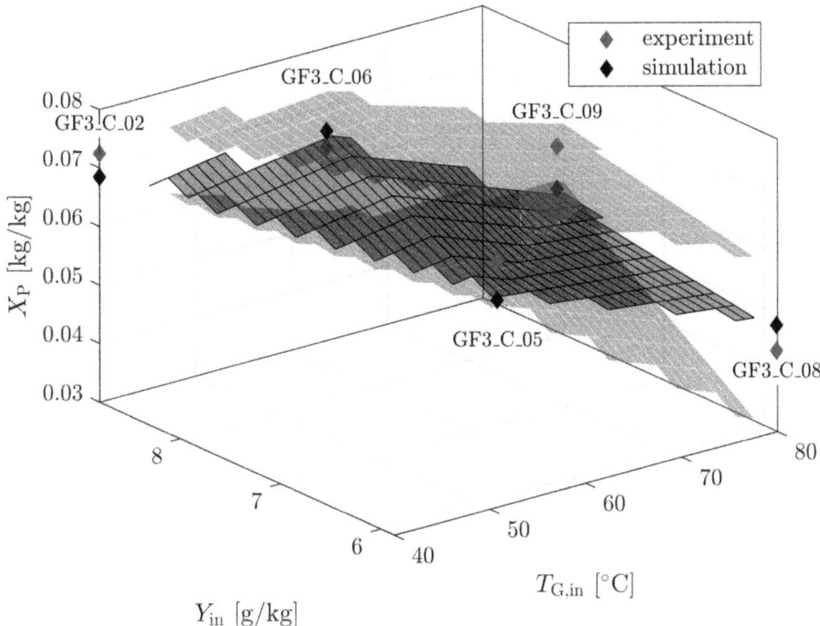


Figure 5.7: Comparison of measured and predicted particle moisture content ($X_P$), plotted against drying temperature ($T_{G,in}$) and inlet moisture content of air ($Y_{in}$) in *GF3* lab-scale dryer for Cellets. Dark gray area designates model predictions; light blue areas indicate the upper and lower boundary of measurement errors. Detailed process conditions of represented data points are given in Table 5.1.

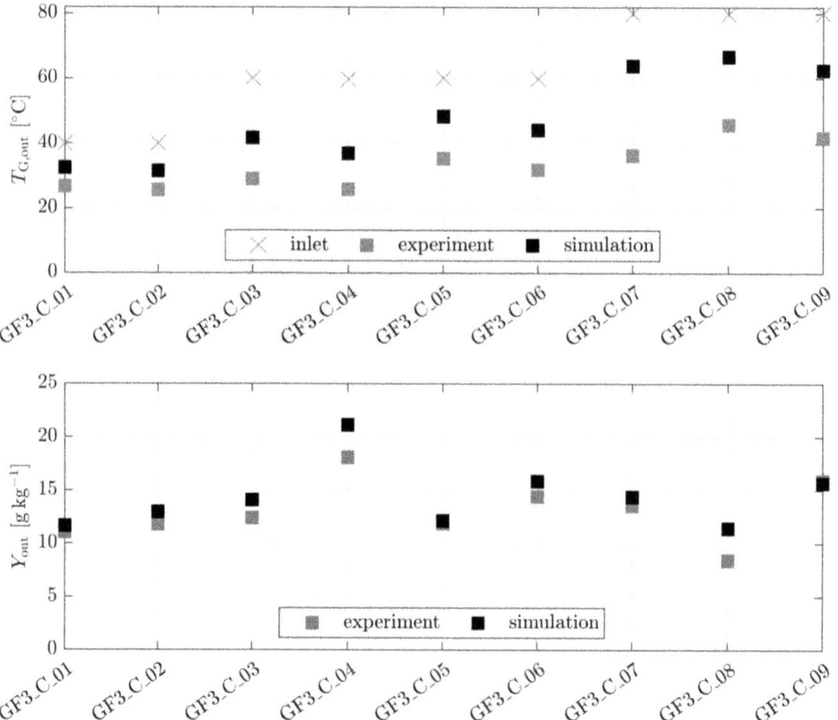


Figure 5.8: Comparison of measured and predicted air temperature ($T_{G,out}$) (top) and moisture content ($Y_{out}$) (bottom), for all validation cases in the *GF3* lab-scale dryer with Cellets; also showing inlet gas temperature. Detailed process conditions of represented data points are given in Table 5.1.

Table 5.1: Overview of investigated process parameters and inlet conditions of experiments and simulations for model validation of cases in the lab-scale *GF3* dryer; including reference names for every experiment.

| Exp. name | Dryer | Material | $T_{G,in}$ [°C] | $u$ [m s$^{-1}$] | $\dot{M}_{P,dry}$ [g s$^{-1}$] | $X_{P,in}$ [kg kg$^{-1}$] | $H_{out}$ [cm] | $\Lambda$ [-] | $Y_{in}$* [g kg$^{-1}$] | $T_{amb}$ [°C] |
|---|---|---|---|---|---|---|---|---|---|---|
| GF3_A_01 | GF3 | $\gamma$-$Al_2O_3$ | 40 | 1.3 | 1.5 | 0.33 | 12 | 0 | 4.64 | 20 |
| GF3_A_02 | GF3 | $\gamma$-$Al_2O_3$ | 60 | 1.3 | 2.0 | 0.50 | 12 | 0 | 6.98 | 21 |
| GF3_A_03 | GF3 | $\gamma$-$Al_2O_3$ | 80 | 1.3 | 1.6 | 0.49 | 12 | 0 | 7.19 | 24 |
| GF3_A_04 | GF3 | $\gamma$-$Al_2O_3$ | 80 | 1.3 | 1.6 | 0.54 | 12 | 0 | 7.79 | 25 |
| GF3_A_05 | GF3 | $\gamma$-$Al_2O_3$ | 80 | 1.3 | 1.9 | 0.45 | 12 | 0 | 10.13 | 21 |
| GF3_A_06 | GF3 | $\gamma$-$Al_2O_3$ | 80 | 1.3 | 1.0 | 0.59 | 7 | 0 | 7.00 | 23 |
| GF3_C_01 | GF3 | Cellets | 40 | 1.1 | 0.7 | 0.21 | 12 | 0 | 8.04 | 23 |
| GF3_C_02 | GF3 | Cellets | 40 | 1.1 | 0.7 | 0.24 | 12 | 0 | 8.81 | 22 |
| GF3_C_03 | GF3 | Cellets | 60 | 0.5 | 0.9 | 0.20 | 12 | 0 | 5.51 | 24 |
| GF3_C_04 | GF3 | Cellets | 60 | 0.5 | 1.2 | 0.23 | 12 | 0 | 10.28 | 22 |
| GF3_C_05 | GF3 | Cellets | 60 | 1.1 | 1.0 | 0.20 | 12 | 0 | 6.76 | 21 |
| GF3_C_06 | GF3 | Cellets | 60 | 1.1 | 1.1 | 0.26 | 12 | 0 | 8.45 | 22 |
| GF3_C_07 | GF3 | Cellets | 80 | 0.5 | 0.5 | 0.24 | 12 | 0 | 7.40 | 22 |
| GF3_C_08 | GF3 | Cellets | 80 | 1.1 | 0.9 | 0.24 | 12 | 0 | 5.86 | 22 |
| GF3_C_09 | GF3 | Cellets | 80 | 1.1 | 1.3 | 0.23 | 12 | 0 | 8.07 | 22 |

* Moisture content of ambient and inlet air are identical ($Y_{in} = Y_{amb}$)

Table 5.2: Overview of investigated process parameters and inlet conditions of experiments and simulations for model validation of cases in the pilot-plant-scale VFB dryer; including reference names for every experiment.

| Exp. name | Dryer | Material | $T_{G,in}$ [°C] | $u$ [$m\,s^{-1}$] | $\dot{M}_{P,dry}$ [$g\,s^{-1}$] | $X_{P,in}$ [$kg\,kg^{-1}$] | $H_{out}$ [cm] | $\Lambda$ [-] | $Y_{in}$* [$g\,kg^{-1}$] | $T_{amb}$ [°C] |
|---|---|---|---|---|---|---|---|---|---|---|
| VFB_C_01 | VFB | Cellets | 40 | 0.3 | 2.7 | 0.24 | 8 | 0.59 | 7.43 | 22 |
| VFB_C_02 | VFB | Cellets | 40 | 0.3 | 5.0 | 0.24 | 8 | 0 | 7.19 | 24 |
| VFB_C_03 | VFB | Cellets | 60 | 0.3 | 3.0 | 0.22 | 8 | 0.59 | 6.07 | 23 |
| VFB_C_04 | VFB | Cellets | 60 | 0.3 | 2.1 | 0.25 | 8 | 0 | 6.77 | 21 |
| VFB_C_05 | VFB | Cellets | 60 | 0.3 | 3.3 | 0.23 | 8 | 0 | 8.81 | 21 |
| VFB_C_06 | VFB | Cellets | 60 | 0.4 | 3.1 | 0.23 | 8 | 0 | 11.80 | 23 |
| VFB_C_07 | VFB | Cellets | 40 | 0.4 | 3.0 | 0.23 | 8 | 0.59 | 10.58 | 22 |
| VFB_C_08 | VFB | Cellets | 60 | 0.4 | 5.3 | 0.23 | 8 | 0 | 10.72 | 22 |
| VFB_C_09 | VFB | Cellets | 60 | 0.4 | 3.2 | 0.24 | 8 | 0.59 | 10.48 | 22 |
| VFB_F_01 | VFB | FCC | 40 | 0.07 | 1.70 | 0.25 | 8 | 0 | 1.06 | 19 |
| VFB_F_02 | VFB | FCC | 40 | 0.07 | 1.70 | 0.24 | 8 | 0.59 | 1.23 | 19 |
| VFB_F_03 | VFB | FCC | 60 | 0.07 | 2.00 | 0.23 | 16 | 0.59 | 1.15 | 19 |
| VFB_F_04 | VFB | FCC | 40 | 0.07 | 1.80 | 0.24 | 16 | 0 | 1.12 | 20 |
| VFB_F_05 | VFB | FCC | 40 | 0.07 | 1.70 | 0.24 | 16 | 0.59 | 1.12 | 20 |
| VFB_F_06 | VFB | FCC | 60 | 0.07 | 0.65 | 0.20 | 16 | 0 | 1.02 | 20 |
| VFB_F_07 | VFB | FCC | 60 | 0.07 | 0.54 | 0.30 | 16 | 1.21 | 1.01 | 20 |

* Moisture content of ambient and inlet air are identical ($Y_{in} = Y_{amb}$)

### 5.1.3 Validation for Dryer Geometries and Vibration

The VFB pilot-plant-scale dryer is used for investigations regarding the influence of vibration on the hydrodynamics and drying kinetics. In contrast to the *GF3* lab-scale dryer, it has a rectangular cross-sectional area, allowing for a comparison of model performance for different dryer geometries and respectively for varying residence time distributions (RTD). For comparison, Cellets particles are investigated in the VFB pilot-plant-scale dryer. Results are illustrated analogously to the previous chapters. The measured RTD characteristics (see chapter 4.2) are included in the simulations via the tank in series model. Process parameters of the investigated cases are listed in Table 5.2. Experimental and simulation results are summarized in Table D.2.

Measurements and model predictions of bed height for different vibration intensities are shown in Figure 5.9. The presented data include different gas flow rates, gas temperatures, vibration intensities and particle mass flow rates. In all cases, model prediction agree well with experimental data, i.e. model prediction lie within the experimental margin of error. Thus, the model delivers accurate predictions of hydrodynamics of vibrated and non-vibrated fluidization of Geldart group B particles.
Vibration has no significant impact on the bed height, in experiments and model prediction alike. This observation is in accordance with previous findings, discussed in chapter 3.3.3. It has been shown that mechanical vibration in the investigated range only reduces $u_{\mathrm{mf}}$, while no further significant effects on the hydrodynamics are observed. Due to the lack of reliable correlations for prediction of $u_{\mathrm{mf}}$ in dependence of moisture content and vibration, the measured $u_{\mathrm{mf}}$ values are used as input parameters in the model.

Figure 5.10 depicts comparison of measured and predicted particle moisture content of Cellets for varied gas flow rate, gas temperature, particle mass flow and vibration intensity. Predictions and validation experiments agree well. Evidently, increased drying temperature results in decreased particle moisture content. Increased particle mass flow leads to increased particle moisture content (see case VFB_C_02). Hence, physical effects are represented correctly by the model, as was achieved in previous validation cases. Additionally, vibration intensity is investigated as another process parameter. Based on the previous findings of almost insignificant impact of vibration on the hydrodynamics, the vibration's impact on drying kinetics and overall model predictions are expected to be minor. This is confirmed by the presented results.
In comparison to investigations in the lab-scale *GF3* dryer, the experimental margin of error is reduced significantly by utilization of a rotary valve for controlling the particle mass flow rate. Simultaneously, this reduces relative deviations between experiments and simulations to a maximum of 16 %. Thereby, the presented results show the highest accuracy of all validation cases, discussed to this point.

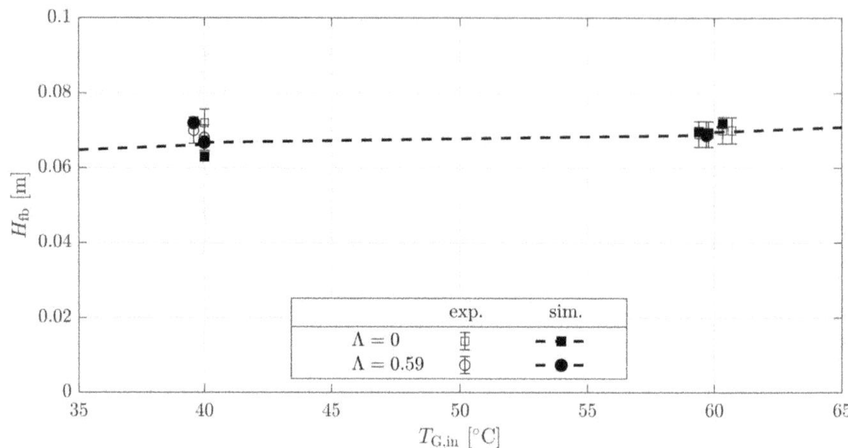


Figure 5.9: Comparison of predicted bed height ($H_{fb}$) and measurements in VFB dryer for Cellets, plotted against gas temperature ($T_{G,in}$) at different gas velocities ($u$) and vibration intensities ($\Lambda$).

Measured and predicted particle temperature are compared in Figure 5.11. Observed trends and dependencies of particle temperature are analogous to the findings in the *GF3* dryer. Accurate predictions are attributed to the model by low relative deviations between model and experiment, not exceeding 4 %. In particular, for low drying temperatures (40 °C) relative deviations are well below 1 %. For higher drying temperatures the model begins to over-predict the particle temperature. Similar behavior is observed for drying experiments in the *GF3* lab-scale dryer for $\gamma$-$Al_2O_3$ particles (see chapter 5.1.1). Analogous, stronger influence of heat transfer from bed to environment may explain this observation. Resulting from the rectangular geometry of the VFB dryer, the ratio of surface area to volume is greater compared to the conical *GF3* dryer. Hence, available heat transfer area is increased and heat transfer rates from particles to the environment become more significant. Therefore, heat transfer from inside the dryer to the environment is more dominant in the rectangular VFB dryer.

Figure 5.12 illustrates measured and predicted temperature and moisture content of the air at the dryer outlet. Similar over-prediction of gas outlet temperature at elevated drying temperatures is observed for simulated outlet air temperature. Likewise, deviations between measured and predicted outlet air moisture content are greater at elevated drying temperatures. Otherwise, outlet air moisture content is predicted accurately by the model.

Despite not considering heat transfer from particles to environment, particle temperature and moisture content are predicted with reliable accuracy. This holds for different dryer geometries as well as cases with and without vibration. The influence of vibration

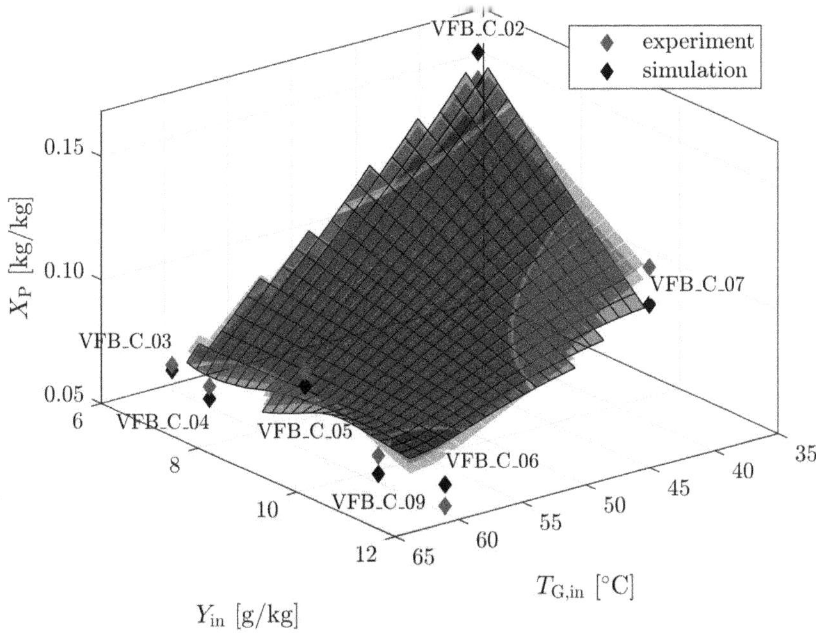


Figure 5.10: Comparison of measured and predicted particle moisture content ($X_P$), plotted against drying temperature ($T_{G,in}$) and inlet moisture content of air ($Y_{in}$) in VFB dryer for Cellets. Dark gray area designates model predictions; light blue areas indicate the upper and lower boundary of measurement errors. Detailed process conditions of represented data points are given in Table 5.2.

is covered by consideration of reduced $u_{mf}$ of the particles in the model. Dependencies of particle residence time distribution on dryer geometry and process conditions are also considered in the model, based on experimental results.

Further optimization of the model is possible, it would however require case specific adjustments or assumptions. Up to this point, fitting parameters have not been used and assumption have been kept to a minimum or have been justified by measurements. Case specific optimizations of the model are analyzed and discussed in combination with sensitivity analyses in chapter 5.2.

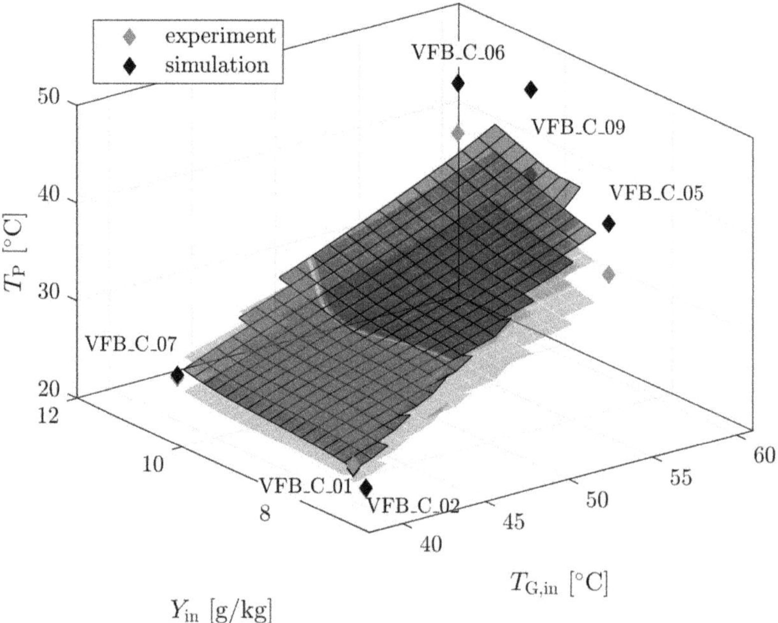


Figure 5.11: Comparison of measured and predicted particle temperature ($T_P$), plotted against drying temperature ($T_{G,in}$) and inlet moisture content of air ($Y_{in}$) in VFB dryer for Cellets. Dark gray area designates model predictions; light red areas indicate the upper and lower boundary of measurement errors. Detailed process conditions of represented data points are given in Table 5.2.

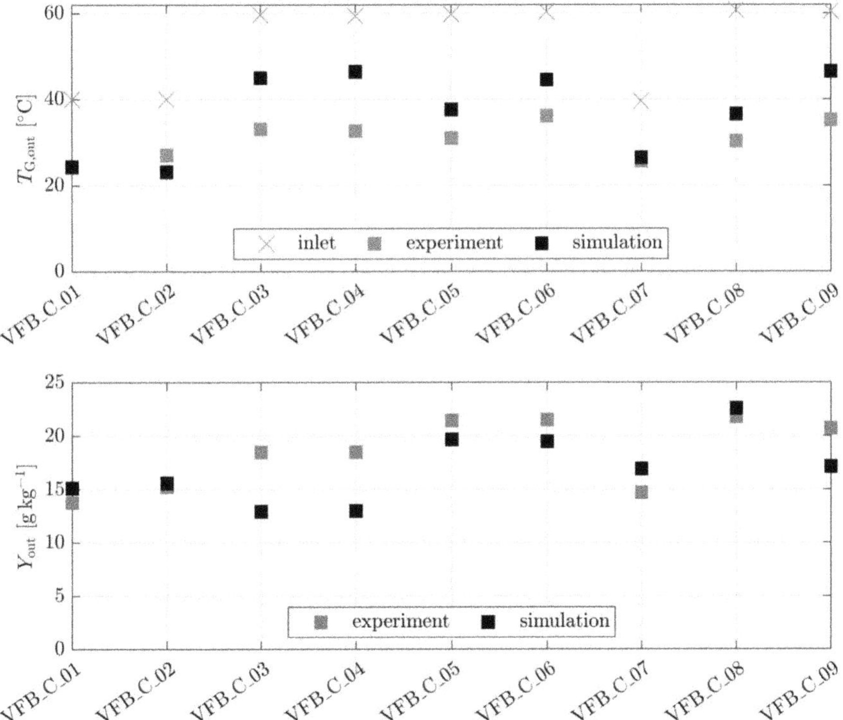


Figure 5.12: Comparison of measured and predicted air temperature ($T_{G,out}$) (top) and moisture content ($Y_{out}$) (bottom), for all validation cases in the VFB pilot-plant-scale dryer with Cellets; also showing inlet gas temperature. Detailed process conditions of represented data points are given in Table 5.2.

### 5.1.4 Validation for Geldart A Particles

FCC catalyst particles are significantly smaller than the Cellets. Literature review (see chapter 2.1.4) indicates that lower particle size leads to increased dispersion, lower Bodenstein number and thus, to RTD behavior closer to CSTR. In combination with the previous observation that the RTDs of Cellets in the VFB only deviate slightly from CSTR behavior, ideally mixed CSTR residence time characteristics are assumed for FCC catalyst in the VFB dryer.

The increased cohesiveness due to lower particle size results in FCC particles adhering to the walls of the dryer during the experiments. Hence, the uncertainty of the optical measurement of bed heights increases, resulting in higher error margin, compared to other investigated powders. Figure 5.13 illustrates comparisons of measured and predicted bed height of FCC catalyst in the VFB dryer for different particle mass flow ($\dot{M}_{\mathrm{P,in}}$) and height of the outflow hole. It shows, that the measurements and model predictions agree reasonably well, i.e. predictions lie within the margin of experimental error in all cases, except one.

Dried pressurized air is used for fluidization of FCC catalyst, because the required gas flow rates are lower than the rotary blower could deliver. The pressurized air shows constant inlet moisture content of approximately 1 g/kg. In contrast to previous validation cases, the FCC catalyst's low particle diameter compels the consideration of heat transfer from particles to dryer walls. Martin (2010) quantified the influence of particle size on the heat transfer coefficient between walls (or immersed surfaces) and particles. Approximately five fold increase of the heat transfer coefficient was reported for particle size reduction from Geldart B and D particles to the size range of the FCC catalyst (Martin, 2010). Also, the VFB dryer has higher ratio of wall area per volume than the *GF3* dryer. This further increases the impact of heat transfer between particles and dryer wall. These arguments are confirmed by simulations with FCC catalyst. In cases with neglected heat transfer between particles and dryer wall, particle temperature is strongly over-predicted while particle moisture content is under-predicted. Consideration of heat transfer between particles and dryer walls leads to much better agreement of model predictions and experiments. In these cases, the wall temperature at the inside of the dryer is assumed to equal ambient temperature.

The validation results are plotted with respect to the influence of particle mass flow $\dot{M}_{\mathrm{P}}$ and inlet moisture content of the particles $X_{\mathrm{P,in}}$. In Figure 5.14 measured and predicted particle moisture content are compared. The comparison shows accurate predictions of particle moisture content, mostly within the experimental margin of error, but never deviating from the experiments by more than 16 %. Physical effects, like reduced particle moisture content for lower particle mass flow and particle inlet moisture content, are accurately predicted by the model. Overall, the influence of vibration, drying temperature, particle mass flow and particle inlet moisture content is therefore well represented qualitatively and quantitatively.

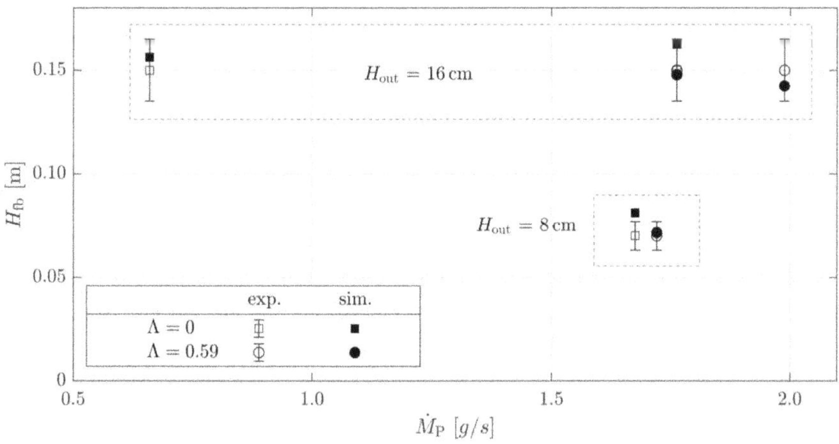


Figure 5.13: Comparison of predicted bed height ($H_{fb}$) and measurements in VFB dryer for FCC catalyst for different outlet heights ($H_{out}$), plotted against particle mass flow rate ($\dot{M}_P$).

Figure 5.15 analogously shows the validation of the model in terms of particle temperature. The predicted particle temperatures never deviate by more than 1 % from the measurements and all lie within the experimental margin of error.

Predicted and measured temperature and moisture content of drying air at the outlet of the dryer are illustrated and compared in Figure 5.16. Simulated gas temperatures are in good agreement with measured values. In contrast, the measured gas moisture content mostly lies below the predicted values. As discussed above, is this deviation likely based on the distance between measurement location and particle bed. Furthermore, fine fractions of the FCC catalyst may have covered the cap of the humidity sensor, and thereby interfered with the measurement. Dried fines, covering the humidity sensors cap, would hinder humid process air from reaching the sensor and thereby explain lower moisture content measurement than expected and predicted by the model.
The presented data leads to the conclusion that the proposed model is able to accurately predict the fluidized bed drying of FCC catalyst (Geldart group A) under consideration of heat transfer between particles and dryer wall. Assuming that the wall temperature equals ambient temperature is a valid approximation in this case, because the VFB dryer is not thermally insulated. This assumption and the general impact of heat transfer between particles and dryer wall is discussed more closely in chapter 5.2.4.

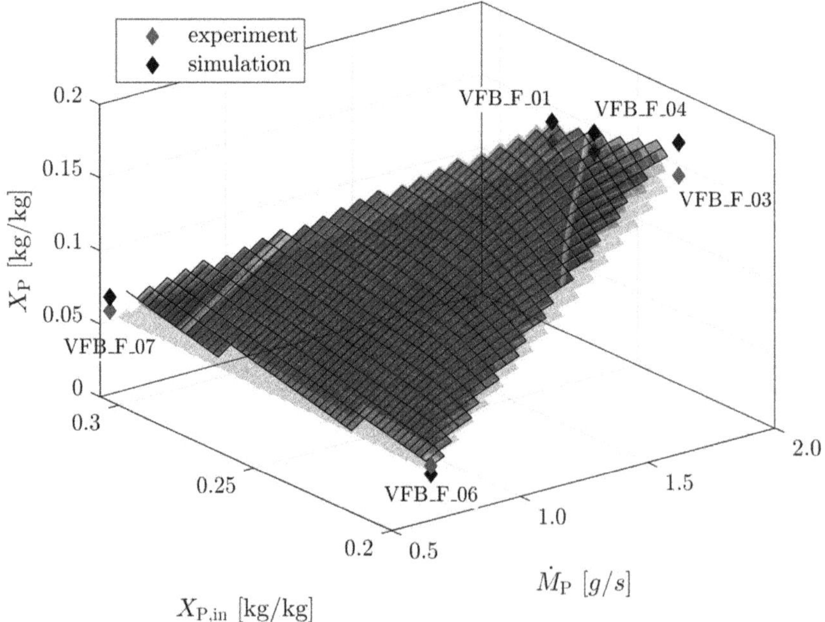


Figure 5.14: Comparison of measured and predicted particle moisture content ($X_P$), plotted against particle mass flow ($\dot{M}_{P,in}$) and particle inlet moisture content ($X_{P,in}$) in the VFB pilot-plant-scale dryer for FCC catalyst. Also drying temperature is varied: $T_{G,in} = 60\,°C$ in cases VFB_F_3, VFB_F_6, VFB_F_7 and $T_{G,in} = 40\,°C$ for the remaining cases. Dark gray area designates model predictions; light blue areas indicate the upper and lower boundary of measurement errors. Detailed process conditions of represented data points are given in Table 5.2.

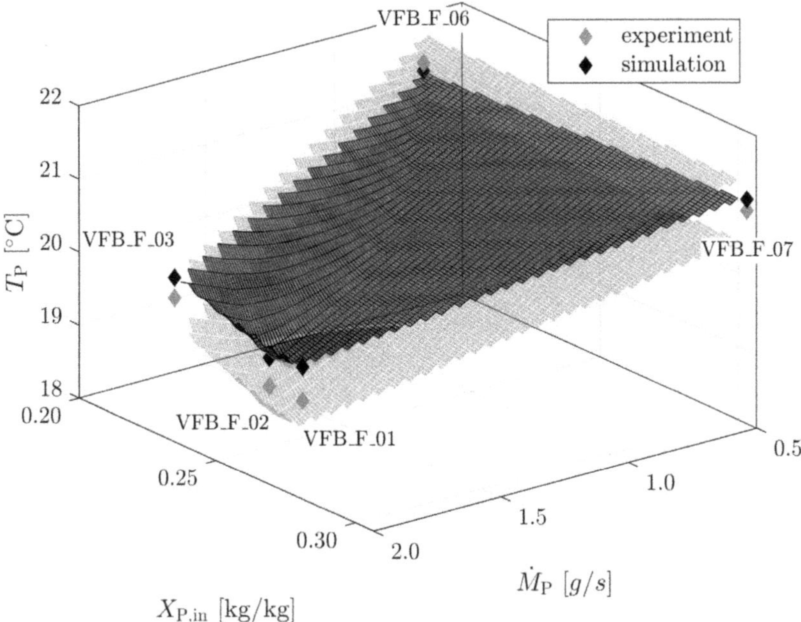


Figure 5.15: Comparison of measured and predicted particle temperature ($T_P$), plotted against particle mass flow ($\dot{M}_{P,in}$) and particle inlet moisture content ($X_{P,in}$) in the VFB pilot-plant-scale dryer for FCC catalyst. Also drying temperature is varied: $T_{G,in} = 60\,°C$ in cases VFB_F_3, VFB_F_6, VFB_F_7 and $T_{G,in} = 40\,°C$ for the remaining cases. Dark gray area designates model predictions; light red areas indicate the upper and lower boundary of measurement errors. Detailed process conditions of represented data points are given in Table 5.2.

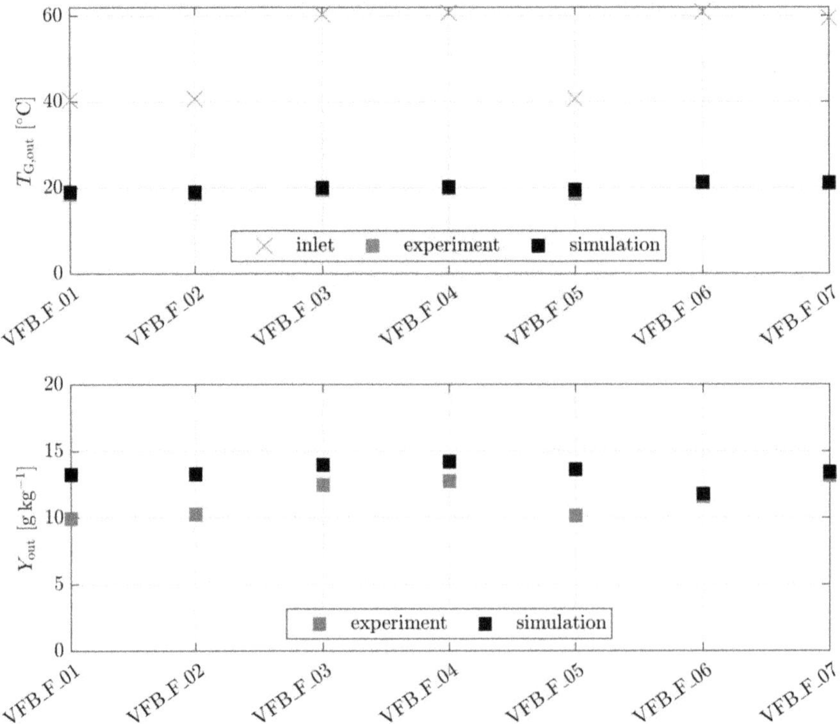


Figure 5.16: Comparison of measured and predicted air temperature ($T_{G,out}$) (top) and moisture content ($Y_{out}$) (bottom), for all validation cases in the VFB pilot-plant-scale dryer with FCC catalyst; also showing inlet gas temperature. Detailed process conditions of represented data points are given in Table 5.2.

## 5.2 Sensitivity Analysis

Reference cases are defined for sensitivity analysis of the model, representing different materials, process conditions and dryer geometries (GF3_A_02, GF3_C_03 and VFB_F_02). In the sensitivity analysis, one parameter is varied around the reference value, while the remaining parameters are unchanged. The reference cases and underlying simulation parameters are listed in Table 5.3. The focus of this sensitivity analysis lies on assumptions made in the development of the model as well as previous applications of those assumptions.

The number of height classes and residence time classes represent the numerical grid of the simulations. Grid independence studies show that simulation results are not effected, when the number of residence time classes is $> 100$ and the number of height classes is $> 50$.

The impact of the cut off residence time value $\tau_{max}$ (see chapter 4.4) is more pronounced. Reduced $\tau_{max}$ leads to more particles with high residence time being excluded from the calculations. Excluding particles with the highest residence time results in increased particle moisture content and decreased particle temperature, as particles with high residence time will be dried further than newer particles. Detailed analysis revealed that the model predictions lie within the experimental margin or error for $\tau_{max} > 90\,\%$. As expected, increasing grid refinement as well as $\tau_{max}$ leads to significant increase in computation time.

The results of the discussed sensitivity analyses of the solver parameters are illustrated in Appendix D (Figure D.1 through D.3). The models sensitivity towards process conditions and operation parameters is topic of the following chapter.

### 5.2.1 Process Parameters

The hydrodynamic model is influenced mainly by the particle properties. They are represented in the model via Geldart group (particle size and density) and the resulting $u_{mf}$. The Geldart group defines the used correlations and parameters for calculation of the bubble volume fraction $\varepsilon_B$. Detailed description of these correlations is given in chapter 4.4.1. In general, larger bubble volume fraction results in smaller portions of gas in direct contact with the particles. Hence, heat and mass transfer rates are reduced between suspension and particles, leading to higher particle moisture content and lower particle temperature. However, these effects are only marginal on the overall model results. I.e. model predictions still lie within the experimental margin of error, when changing the hydrodynamic correlations (see Figure D.4).

Furthermore, $u_{mf}$ is used in the calculation of bubble volume fraction and bed height. Negligible sensitivity of the overall model predictions is observed for variation of $u_{mf}$ (see Figure D.5).

Table 5.3: Simulation parameters of reference cases for sensitivity analysis.

| Parameter | GF3_A_02 | GF3_C_03 | VFB_F_02 |
|---|---|---|---|
| Anderson acceleration factor | 10 | 10 | 10 |
| Number of height classes | 100 | 100 | 100 |
| Number of residence time classes | 400 | 400 | 400 |
| Maximum residence time $\tau_{max}$ | 0.95 | 0.99 | 0.99 |
| Geldart group | D | B | A |
| $u_{mf}$ [m/s] | 0.53 | 0.2 | 0.014 |
| $M_P$ [kg] | 1.7 | 1.1 | 5.8 |
| $\dot{M}_P$ [g/s] | 2.05 | 1.13 | 1.72 |
| $T_{P,in}$ [K] | 294 | 295 | 293 |
| $X_{P,in}$ [kg/kg] | 0.50 | 0.26 | 0.24 |
| $\dot{M}_G$ [kg/s] | 0.033 | 0.03 | 0.01 |
| $T_{G,in}$ [K] | 333 | 333 | 314 |
| $Y_{in}$ [g/kg] | 6.98 | 8.45 | 1.23 |
| $\Lambda$ [−] | 0 | 0 | 0.59 |
| $K$ [−] | 1 | 1 | 1 |
| $T_W$ [K] | n.a. | n.a. | 293 |
| $H_{NTU,SB=1}$ [m] | 0.05 | 0.05 | 0.05 |

n.a. = not applied

As described in chapter 4.3.1, vibration has negligible influence on the drying kinetics of the investigated particles. This is represented by the model, which consequently shows negligible sensitivity with respect to changing vibration frequencies and amplitudes. This is to be expected because the investigated particles do not agglomerate due to interparticle forces. In cases, where agglomeration occurs, vibration is expected to reduce the agglomerate size and thereby influence the particle size and thereby the hydrodynamics and indirectly the drying kinetics.
The model shows higher sensitivity towards process parameters that have direct impact on heat and mass transfer. The inlet temperature of the particles $T_{P,in}$ is such a parameter. Given that $T_{P,in}$ is below the wet bulb temperature, its influence is negligible. In case $T_{P,in}$ exceeds the wet bulb temperature, the impact becomes more dominant and leads to reduced particle moisture content and increased particle temperature (see Figure D.6). In comparison to other operation parameters, $T_{P,in}$ shows the lowest impact on the simulation results.
The parameters with the highest impact on the overall results are particle inlet moisture content and mass flow rate, as well as inlet temperature and moisture content of the drying air. These process parameters were thoroughly discussed during the model validation. It was shown above that the high sensitivity of the model is founded on physical effects and confirmed by experiments (see chapter 5.1). Thus, they are not discussed here again. Instead the focus shall lie upon the model's sensitivity towards underlying assumptions in the following sections.

### 5.2.2 Model Assumptions

Underlying assumption of the model were made in analogy to previos works by Alaathar (2017) and Burgschweiger (2000), introduced and discussed in chapter 4. Changes in particle size due to agglomeration, attrition or entrainment are neglected.
Visual observation of the particle flow at the inlet and outlet as well as the freeboard showed no indications of neither agglomerates nor dust formation during the experiments for Cellets and $\gamma$-$Al_2O_3$ particles. Only free flowing primary particles were observed. Measurement of the particle size at the inlet and outlet could not elucidate potential formation of temporary agglomerates. Any additional mechanical stress, e.g. from sample taking, transportation or filling material into the *CAMSIZER XT*, or any other external analysis device, is certain to break possible liquid bridges and counteract other cohesive forces. Hence, temporarily formed agglomerates could not be detected. The formation of permanent material bridges is eliminated by only investigating particles, insoluble in water, and using demineralized water in all experiments.
Additionally, the particle size distributions of the materials were measured before and after the experiments. Two to three drying experiments were performed per day for validation purposes. These measurements showed changes in Sauter mean diameter in the sub-micrometer range. Comparison of the measured particle size distributions showed almost perfect overlap. Hence, these plots are not shown here.
Based on the discussed observations, the model assumption of unchanged particle size distributions is confirmed to be valid for the investigated materials and process conditions. Furthermore, this conclusion is supported by the high accuracy of the model predictions, shown in chapter 5.1.

### 5.2.3 Residence Time Distribution

The model's sensitivity towards the residence time distribution (RTD) is analyzed in this section. The theoretical tank number $K$ according to the tank in series number is varied in 10 steps, from ideally mixed CSTR ($K = 1$) to conditions close to ideal plug flow reactor ($K = 100$). Resulting model predictions are illustrated for all three reference cases in Figure 5.17 regarding average particle moisture content and particle temperature at the dryer outlet.
The main impact of increasing tank number is the reduction of dispersion, meaning the mixing of wet particles (newly entering the bed) with particles with higher residence time and thus, already dried. Compared to case GF3_C_03 (Figure 5.17 middle), the sensitivity analysis based on cases GF3_A_02 and VFB_F_02 (Figure 5.17 (top) and (bottom), respectively) show different trends for particle moisture content and temperature with increasing theoretical tank number. This is based on the different process conditions. In case GF3_C_03, the particles are in the second drying period, whereas the other two cases represent particles in the first drying period.
In the second drying period, particles are dried to the equilibrium moisture content and

particle temperature approaches the inlet temperature of the drying gas. This is evident for case GF3_C_03 for increasing theoretical tank number. For low tank numbers, close to an ideally mixed CSTR, new and wet particles are mixed with the particles with high residence time, resulting in average particle moisture content higher than the equilibrium moisture content and lower average particle temperature than the gas temperature.

In cases, where the particles are in the first drying period all along, the trend is different. Here, particle temperature equals the wet bulb temperature given sufficient residence time. Hence, the particle temperature is not affected as significantly by increasing tank numbers in cases GF3_A_02 and VFB_F_02. With increasing residence time, the particles are dried further. Therefore, reduced mixing, with increased tank number under constant mean residence time results in lower particle moisture content.

The described and observed dependencies are expected and thereby correctly accounted for by the model. Especially, when particles are processed in the second drying period, residence time characteristics play an important role and are crucial to the model predictions.

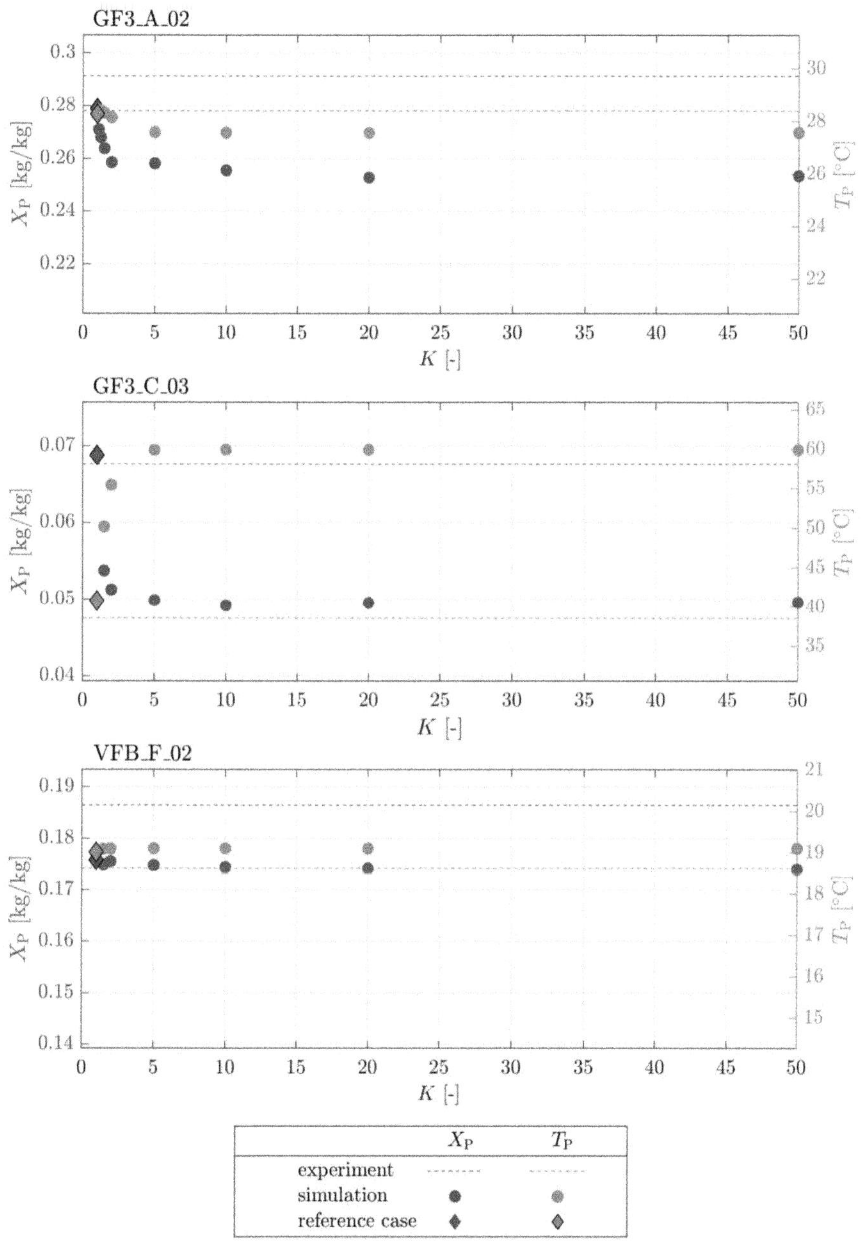


Figure 5.17: Sensitivity analysis of model predictions with regard to theoretical number of tank according to the tank in series model ($K$); showing impact on particle moisture content ($X_P$) and particle temperature ($T_P$) for reference cases GF3_A_02 (top), GF3_C_03 (middle) and VFB_F_02 (bottom).

### 5.2.4 Heat Transfer - Dryer to Environment

During validations (chapter 5.1) it is observed that the model over-predicts particle temperature at elevated drying temperatures as well as small particles of Geldart group A. The most likely explanation is the increasing influence of heat transfer from the dryer to the environment at high drying temperatures and small particle diameters. This aspect is analyzed in this chapter.

As discussed in chapter 4.4.2, heat transfer is considered in the model in form of heat transfer from drying chamber to dryer wall (see schematic in Figure 4.18). Therein, the wall temperature at the inside of the dryer needs to be known or assumed. In case of ideal heat transfer from drying chamber to environment, the wall temperature would equal ambient temperature. For real cases, the temperature of the inside wall will be higher than ambient temperature. This is based on resistances to heat transfer from particle bed to the wall, heat conductivity through the wall and possible insulation as well as heat transfer from the outside wall to the environment.

The utilized fluidized bed dryers are not equipped with thermal insulation. Their walls consist of stainless steel and glass. Thus, the temperature of the inner dryer wall will be slightly higher than ambient temperature. For the first step of sensitivity analysis, it is therefore assumed that the wall temperature equals ambient temperature. These cases, including heat transfer from dryer to environment are compared to the validation cases from the VFB dryer with Cellets (heat transfer to environment neglected, see chapter 5.1.3). The resulting particle temperatures and gas temperatures are further compared to experimental values and shown in Figure 5.18.

The results confirm the hypothesis regarding the influence of heat transfer between particles and dryer wall $\dot{Q}_{\mathrm{PW}}$. For drying temperatures of 40 °C, the model shows only negligible sensitivity and the predictions agree well with experimental data. As drying temperature is increased, the sensitivity towards $\dot{Q}_{\mathrm{PW}}$ grows. Consideration of $\dot{Q}_{\mathrm{PW}}$ at drying temperatures around 60 °C results in reduction of predicted gas temperature as well as predicted particle temperature. Predictions of both parameters are significantly closer to experimental values for high drying temperatures. The impact on resulting particle moisture content is only minor in the investigated range and is therefore not shown.

The presented data show that the assumption of dryer wall temperature being equal to ambient temperature is a reasonable assumption in the investigated cases. However, the exact wall temperature of the dryer could not be measured in the experiments. Furthermore, it is a case specific parameter, that will vary from dryer to dryer. It may even differ, for different parts of the one dryer, due to different wall materials or insulation. Considering all this, the option of considering $\dot{Q}_{\mathrm{PW}}$ in the simulations is required for elevated drying temperatures as well as small particles. However, the exact value of the wall temperature, remains an assumption and may be seen as a fitting parameter.

Another potential use of the $\dot{Q}_{\mathrm{PW}}$-feature is representation of additional heat sources, or

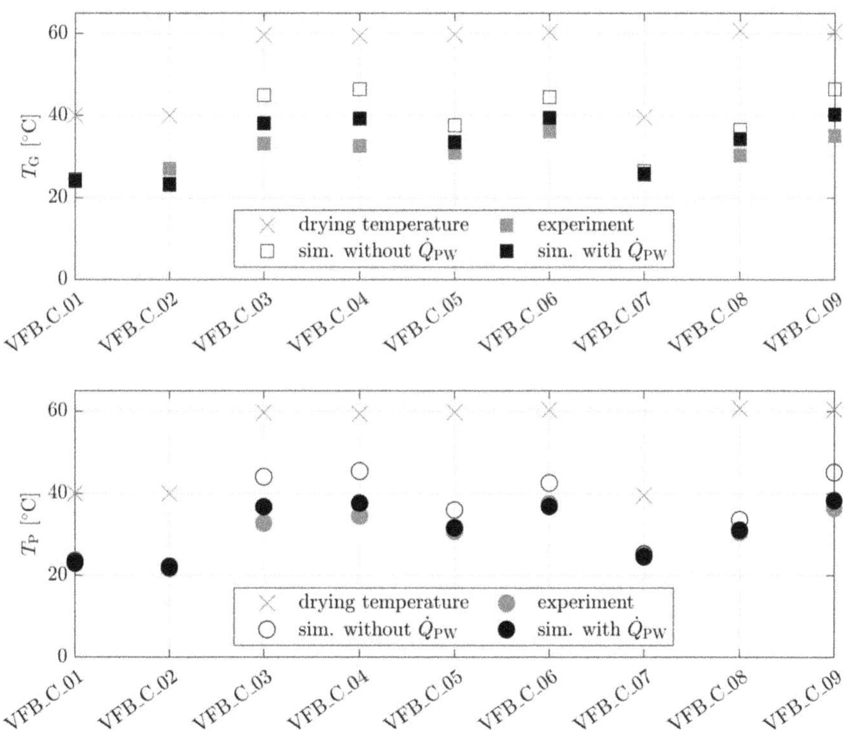


Figure 5.18: Sensitivity analysis of heat transfer between particle and walls ($\dot{Q}_{PW}$) on predicted outlet gas temperature ($T_G$) (top) and particle temperature ($T_P$) (bottom) validation cases in VFB dryer with Cellets, also showing respective drying temperature and measured temperatures. Detailed process parameters are listed in Table 5.2.

sinks, via dryer walls or internal heat transfer surfaces. These cases are not uncommon, but require specific validation before reliable application within the proposed model. An exemplary sensitivity study is illustrated in Figure D.7.

Additionally, heat transfer from drying gas to dryer wall could be considered to increase the models accuracy. This option is pre-implemented in the model. However, comparison of heat transfer coefficients from gas to wall and particles to wall identified negligible impact of heat transfer from gas to walls, as the respective heat transfer coefficients only equate a few percent of the transfer coefficient between particles and wall in the investigated parameter range. Thus, and for reasons of simulation stability, heat transfer from drying gas to dryer walls is neglected.

### 5.2.5 Number of Transfer Units

Another assumption, made in the model, is that the number of transfer units between bubbles and suspension phase $NTU_{SB}$ equals 1 at a bed height of 5 cm. Based on this, the heat transfer coefficient between bubbles and suspension $\beta_{SB}$ is calculated:

$$\beta_{SB} = \frac{NTU_{SB} \cdot \dot{M}_G}{\rho_G \cdot A_{SB}} \quad \text{with: } NTU_{SB} = \frac{H_{fb}}{0.05m}. \tag{5.1}$$

More details are given in Appendix C.3. Theoretical limits of the number of transfer units are $NTU_{SB} = 0$ (meaning inactive bypass, i.e. no mass transfer from suspension to bubble phase) and $NTU_{SB} \to \infty$ (meaning no bypass, i.e. instant and perfect mass transfer between suspension and bubble phase) (Groenewold and Tsotsas, 1997).
The assumption that $NTU_{SB}=1$ at $H_{fb}=0.05$ m was used in preceding versions of this model (Alaathar, 2017, Alaathar et al., 2013, Burgschweiger, 2000, Burgschweiger and Tsotsas, 2002). It was simply used, but did not receive further attention or validation in the literature so far. The assumption was introduced by Groenewold and Tsotsas (1997), who stated that overall drying kinetics in fluidized beds are dominated by the mass transfer between particles and suspension gas. The overall drying kinetics were deemed less sensitive towards mass transfer between bubbles and suspension (Groenewold and Tsotsas, 1997). This assumption seems to be valid, because its application in previous models of continuous fluidized bed dryers delivered accurate results (Alaathar, 2017, Alaathar et al., 2013, Burgschweiger, 2000, Burgschweiger and Tsotsas, 2002). Values for $NTU_{SB}$ for fluidized bed applications, were reported to vary mostly from 0.01 to 100 (Groenewold and Tsotsas, 1997). Bubble characteristics and thereby particle properties (i.e. Geldart group) influence the value of $NTU_{SB}$. Clear dependencies on Reynolds number or bed height have not been identified (Groenewold and Tsotsas, 1997).
The assumption that $NTU_{SB}=1$ at $H_{fb}=0.05$ m results in values for $NTU_{SB}$ between 1.36 to 2.6 in the validation cases of this work (chapter 5.1). Particles of three different Geldart groups are investigated in this work. Considering that $NTU_{SB}$ values are two orders of magnitude higher or lower were found valid for fluidized beds (Groenewold and Tsotsas, 1997), the height, at which $NTU_{SB}=1$ might be treated as a fitting parameter in the model.
The model's sensitivity with regard to $NTU_{SB}$ is investigated for reference case GF3_A_02. The results are shown in Figure 5.19 (top). The original assumption of $NTU_{SB} = 1$ at $H_{fb}=0.05$ m results in good agreement with the experimental values for particle moisture content and temperature, alike. Model predictions show high sensitivity towards variation of $NTU_{SB}$, which is directly proportional to $\beta_{SB}$. Decreasing $NTU_{SB}$ leads to strong over-prediction of particle moisture content and slight under-prediction of particle temperature. On the other hand, increasing $NTU_{SB}$ has only minor impact on the model predictions. Higher values of $NTU_{SB}$ negate the effect of bubbles acting as bypass and thereby, result in higher drying rates. In fluidized beds of Geldart group D particles, bubbles rise slower than the suspension gas. Hence, the suspension gas travels through those bubbles, resulting in (almost) ideal heat and mass transfer between the two phases and thereby, leading to low bypass effect of the bubbles (Chen et al., 2017). Consequently, lower values for $NTU_{SB}$ are unreasonable for Geldart group D particles.

Similar trends are observed for the reference case with Geldart B particles (GF3_C_03), shown in Figure 5.19 (middle). However, the influence of reduced $NTU_{SB}$ is even more pronounced than for Geldart D particles. In contrast to Geldart D, bubbles in Geldart A and B particles rise faster through the bed than the suspension gas. Therefore, the bypass effect of bubbles is more pronounced than for Geldart D particles (Kunii et al., 2013) and the impact of variation in $NTU_{SB}$ are more pronounced for Geldart B particles.

For reference case FCC_VFB_02 (bottom plot in Figure 5.19), the bubble volume fraction is significantly lower than for Geldart B and D particles (0.09 compared to 0.70 and 0.26 averaged over the entire bed height, for Geldart B and D particles respectively). Therefore, the impact of changes in $NTU_{SB}$ are comparably low for the cases with FCC catalyst. In all cases, the original assumption of $NTU_{SB} = 1$ at $H_{fb} = 0.05\,m$ is confirmed to deliver accurate model predictions in the investigated cases, by showing good agreement of predicted particle temperature and moisture content with experimental data.

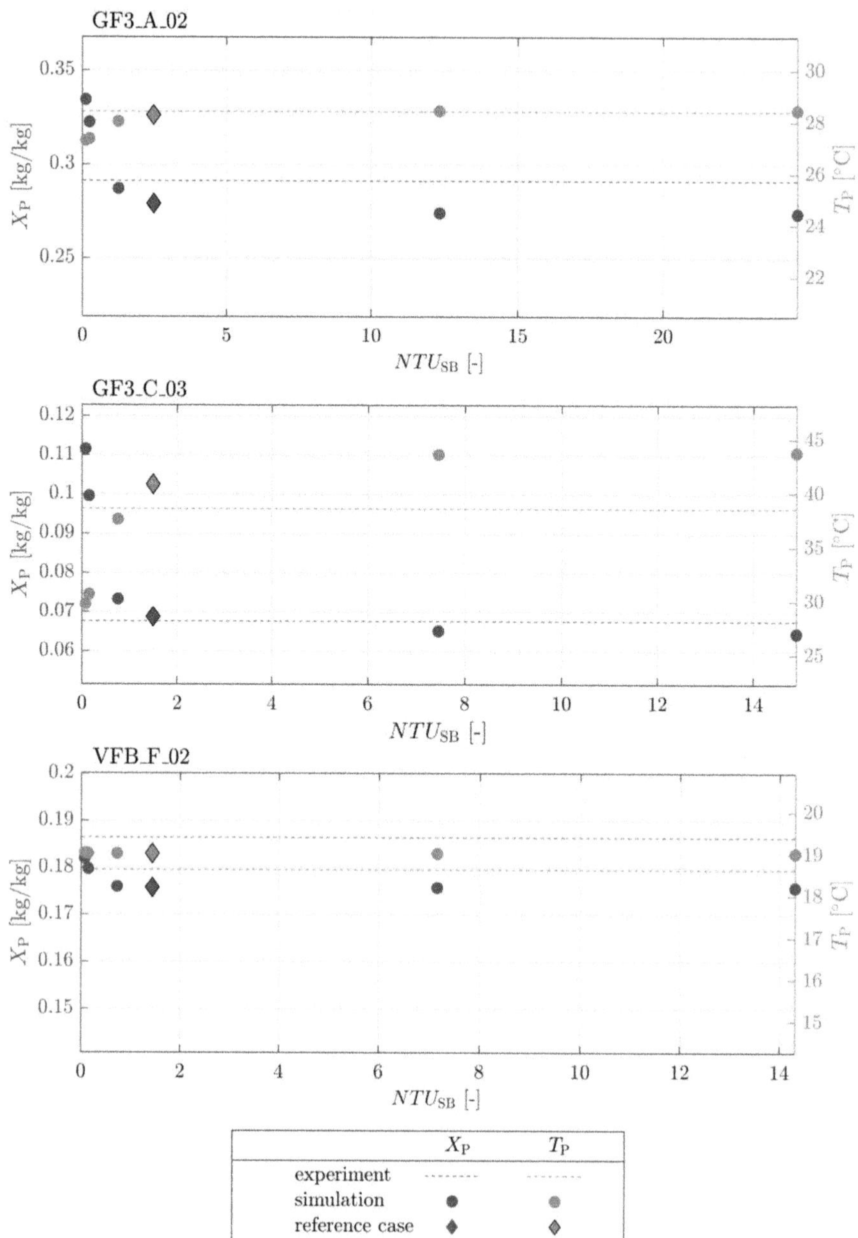


Figure 5.19: Sensitivity analysis of model predictions with regard to the number of transfer units between suspension and bubble phase ($NTU_{SB}$); showing impact on particle moisture content ($X_P$) and particle temperature ($T_P$) for reference cases GF3_A_02 (top), GF3_C_03 (middle) and VFB_F_02 (bottom).

## 5.3 Application

The proposed model is applied to the example of continuous milk powder production on small industrial scale. Key components of the milk powder production facility are a co-current spray dryer, with a cylindrical fluidized bed dryer underneath it. The fluidized bed measures 1.2 m in diameter. Particles, leaving the spray dryer, fall directly into the fluidized bed. Dried particles leave the cylindrical bed via overflow weir and rotary valve. Hereby, the rotary valve minimizes inflow of air from subsequent process steps. A simplified flowchart of the production plant is depicted in Figure. 5.20, including measurement locations of gas flow rate, gas temperature and humidity as well as particle sampling for determination of particle moisture content. The investigated dryer is operated by *TetraPak CPS*, who kindly provided the experimental data.
Simulations are performed, using the settings listed in Table.5.4. Process conditions are presented in Table. 5.5. Solver parameters are chosen analogous to previously discussed validation cases. Also, the temperature of the inside of the dryer wall is assumed to equal ambient temperature. Based on the cylindrical geometry of the dryer, ideal CSTR behavior is used for simulations, i.e. theoretical tank number $K = 1$. As the particle inlet temperature cannot be measured during experiments, values are estimated based on water mass balance around the spray dryer. The critical moisture content of WMP is reported to be well above the measured inlet conditions (Pisecký, 2012) (see Table. 5.5). Thus, the particles have been in the second drying period while still in the spray dryer. Due to high temperatures in the spray dryer, it is assumed that the particles enter the fluidized bed dryer with approximately the same temperature as they have at the outlet of the fluidized bed. According to Langrish and Kockel (2001), the material specific drying curve of WMP is assumed to have linear shape in the second drying period.
Comparisons of experimental data and model predictions are illustrated in Figure.5.21. Particle moisture content and particle temperature (two top plots) are predicted by the model with high accuracy for all investigated parameter combinations. Likewise, high accuracy of model predictions are evident for outlet air moisture content and temperature (two bottom plots in Figure.5.21).
The presented data confirm applicability of the developed model for industrial scale fluidized bed drying of WMP.

Table 5.4: Simulation parameters of industrial milk powder production.

| Parameter | TP cases |
|---|---|
| Number of height classes | 100 |
| Number of residence time classes | 400 |
| Maximum residence time $\tau_{\max}$ | 0.95 |
| Geldart group | C |
| $u_{\mathrm{mf}}$ [m/s] | 0.2 |
| $\Lambda$ [-] | 0 |
| $K$ [-] | 1 |
| $T_{\mathrm{W}}$ [K] | $T_{\mathrm{amb}}$ |
| $H_{\mathrm{NTU,SB}=1}$ [m] | 0.05 |

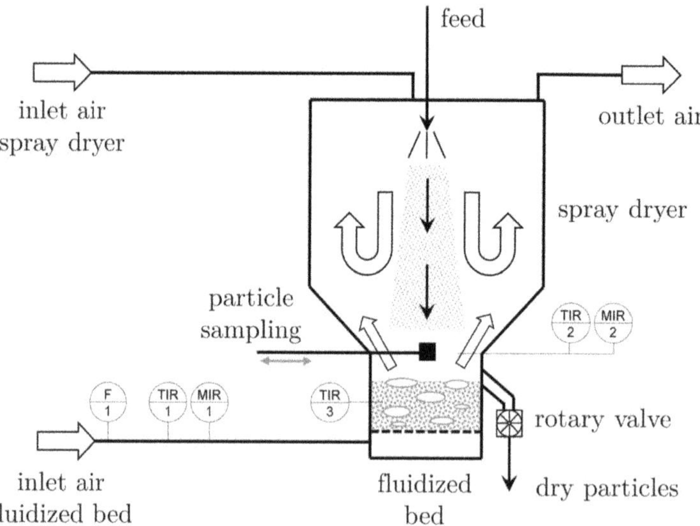


Figure 5.20: Schematic flowchart of small industrial scale whole milk powder production plant, consisting of spray dryer and cylindrical fluidized bed dryer (diameter: 1.2 m). Further showing relevant measurement locations.

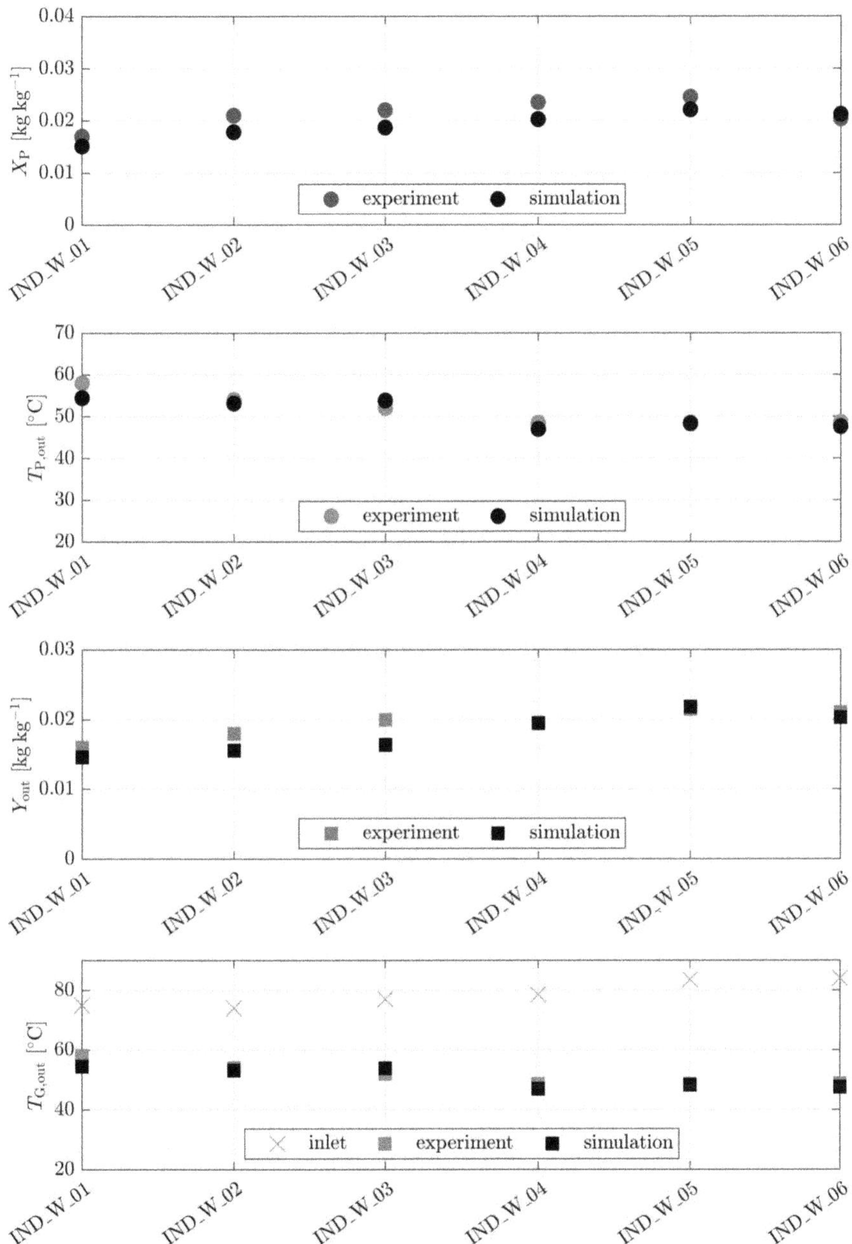


Figure 5.21: Comparison of measured and predicted particle moisture content ($X_{\text{out}}$), particle temperature ($T_{\text{P}}$), air moisture content ($Y_{\text{out}}$) and air temperature ($T_{\text{G,out}}$) (from top to bottom); for test cases on small industrial scale production of whole milk powder; also showing inlet gas temperature. Detailed process conditions of represented data points are given in Table. 5.5.

Table 5.5: Overview of investigated process parameters and inlet conditions of experiments and simulations for model application on small scale production of whole milk powder; including reference names for every experiment.

| Exp. name | Dryer | Material | $T_{G,in}$ [°C] | $u$ [m s$^{-1}$] | $\dot{M}_{P,dry}$ [kg s$^{-1}$] | $X_{P,in}$ [kg kg$^{-1}$] | $H_{out}$ [cm] | $\Lambda$ [-] | $Y_{in}$* [g kg$^{-1}$] | $T_{amb}$ [°C] |
|---|---|---|---|---|---|---|---|---|---|---|
| TP_W_01 | TP | WMP | 75 | 0.72 | 0.45 | 0.060 | 32 | 0 | 9.4 | 28 |
| TP_W_02 | TP | WMP | 74 | 0.74 | 0.56 | 0.071 | 32 | 0 | 9.8 | 29 |
| TP_W_03 | TP | WMP | 77 | 0.69 | 0.52 | 0.074 | 32 | 0 | 9.8 | 30 |
| TP_W_04 | TP | WMP | 79 | 0.69 | 0.24 | 0.113 | 32 | 0 | 7.7 | 27 |
| TP_W_05 | TP | WMP | 84 | 0.69 | 0.23 | 0.125 | 32 | 0 | 7.9 | 29 |
| TP_W_06 | TP | WMP | 84 | 0.71 | 0.19 | 0.133 | 32 | 0 | 6.7 | 26 |

* Moisture content of ambient and inlet air are identical ($Y_{in} = Y_{amb}$)

# 6

# Conclusion

In this thesis, fluidized bed drying of particulate solids is investigated. Special focus lies on the impact of mechanical vibration of fluidized bed dryers. Accurate modeling of the drying process, including the influence of vibration, is the goal of this work. A comprehensive semi-empirical model is developed and implemented in the open-source flowsheet simulation framework DYSSOL and validated with experimental data. Experimental investigations of hydrodynamics and drying kinetics in fluidized beds are performed for model development as well as validation and testing purposes.

The hydrodynamics, and thereby drying kinetics, in fluidized beds depend on process parameters as well as particle properties. Fluidized beds have been investigated extensively in the past. Hence, existing modeling approaches and correlations are reviewed, suitable correlations are chosen for this work and missing or insufficiently examined areas are identified. Thereof, the lack of comprehensive descriptions of fluidized bed hydrodynamics of fine and cohesive particles as well as the influence of vibration is tackled in this thesis.

A custom-built vibrated fluidized bed dryer is designed and constructed for comprehensive investigations of fluidized bed hydrodynamics and drying kinetics. Based on experimental investigations, a set of semi-empirical correlations is developed that accurately models the influence of vibration on the hydrodynamics on small and cohesive particles. This new model is based on established correlations. It accounts for mechanical vibration in a form that, when no vibration is applied, the established correlations are unchanged. Thereby, model validity for conventional and vibrated fluidized bed hydrodynamics is ensured.

Additionally, particle residence time distribution plays an important role in all continuously operated fluidized bed applications. Comprehensive investigation of particle residence time characteristics and the impact of relevant process parameters are investigated experimentally. The resulting dependencies, especially regarding the impact of vibration, confirm and extend reported findings from literature. These results are included in the fluidized bed drying model.

Drying kinetics in fluidized bed models are best described by material specific drying curves. This way, validity over a wide range of process parameters is ensured. Furthermore, comparibility to applications in other dryers is given. For this purpose, the Reaction Engineering Approach (REA) is used in this work. Based on batch-wise fluidized bed drying experiments, the validity of the REA for fluidized bed applications is proved. Consequently, material specific drying curves are determined and can now be expressed independently of the critical moisture content.
The model is implemented in the novel flowsheet simulation framework DYSSOL. Continuously operated fluidized bed drying under steady-state conditions is the focus of this thesis. Comprehensive validation experiments are performed for particles of different Geldart groups, different dryer geometries and a range of process parameters, including mechanical vibration. Comparison of model predictions with experimental data attributes high accuracy of predicted particle and gas properties. The influence of varying process parameters is modeled correctly.
Sensitivity analysis reveals comparably low influence of fluidized bed hydrodynamics on overall model results. Main influencing factors are drying temperature, gas velocity and feed mass flow rate. Further sensitivity analyses are conducted to identify potential weaknesses in underlying model assumptions. Hereby, the validity of underlying assumptions is confirmed and potential optimization parameters for different applications are identified.
Finally, the model is applied successfully on small-scale industrial milk powder drying, underlining the model's accuracy and potential for wide ranging industrial application.

The proposed model is unprecedented in terms of range of process parameters, range of particle properties and dryer geometries, tested and found valid for. Additionally, it is implemented in an open-source flowsheet simulation framework. Future steps could be the combination of this fluidized bed drying model with other unit operations towards the modeling of entire, inter-connected drying process chains. Also, investigation and consideration of entrainment from the dryer and combination with potential recycle streams could prove useful in that regard. Implementation of time-dynamic dependencies could further increase the model's value.

# Appendix A

# Measurement Equipment

**Temperature Sensors**

| Name | Sensor Type | Location |
|---|---|---|
| TIRC1_0 | thermocouple type K | orifice plate |
| TIRC1_1 | thermocouple type K | inlet windbox |
| TIRC1_2 | PT100 | inside the bed |
| TIRC1_3 | PT100 | inside the bed |
| TIRC1_4 | PT100 | inside the bed |
| TIRC1_5 | PT100 | inside the bed |
| TIRC1_6 | PT100 | free board |
| TIRC1_7 | PT100 | ambient |

**Combined Humidity and Temperature Sensors**

| Name | Sensor Type | Location |
|---|---|---|
| HIR1 / HTIR1_1 | SZKA.0H.F182.313.034 | inlet air conditioning unit |
| HIR2 / HTIR1_2 | AWK2.00 F159.F00.00 | inlet windbox |
| HIR3 / HTIR1_3 | AWK2.00 F159.F00.00 | outlet expansion zone |

**Pressure Sensors**

| Name | Sensor Type | Range [mbar] | Location |
|---|---|---|---|
| PDIR1_0 | HCXM350D6H | 0...350 | $\Delta p$ distributor |
| PDIR1_1 | HCXM350D6H | 0...350 | $\Delta p$ distributor and bed |
| PDIR1_2 | HCXM350D6H | 0...350 | $\Delta p$ bed |
| PDIR1_3 | HCXM350D6H | 0...350 | $\Delta p$ windbox to ambient |
| PDIR1_4 | HCXM350D6H | 0...350 | backup sensor |
| PDIR1_5 | HCXM350D6H | 0...350 | $p$ before orifice plate |
| PDIR1_6 | HCXM10D6H | 0...10 | $\Delta p$ orifice plate |
| PDIR2_0 | HCXM100D6H | 0...100 | $\Delta p$ bed increment |
| PDIR2_1 | HCXM100D6H | 0...100 | $\Delta p$ bed increment |
| PDIR2_2 | HCXM100D6H | 0...100 | $\Delta p$ bed increment |
| PDIR2_3 | HCXM20D6H | 0...20 | $\Delta p$ bed increment |
| PDIR2_4 | HCXM20D6H | 0...20 | $\Delta p$ bed increment |
| PDIR2_5 | HCXM20D6H | 0...20 | $\Delta p$ bed increment |

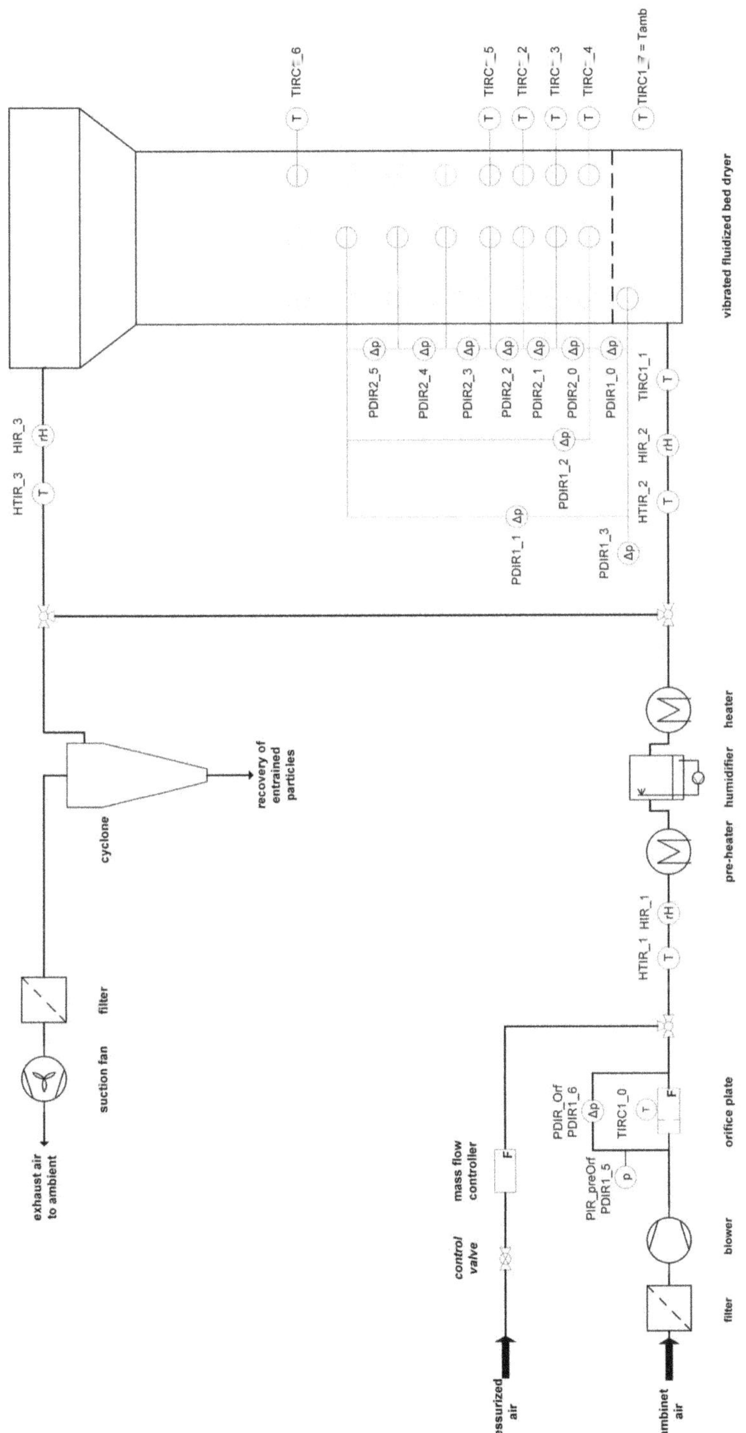


Figure A.1: Flowchart of the periphery of the VFB dryer, including installed measurement equipment and their respective positions.

# Appendix B

# Supplementary Drying Kinetics Data

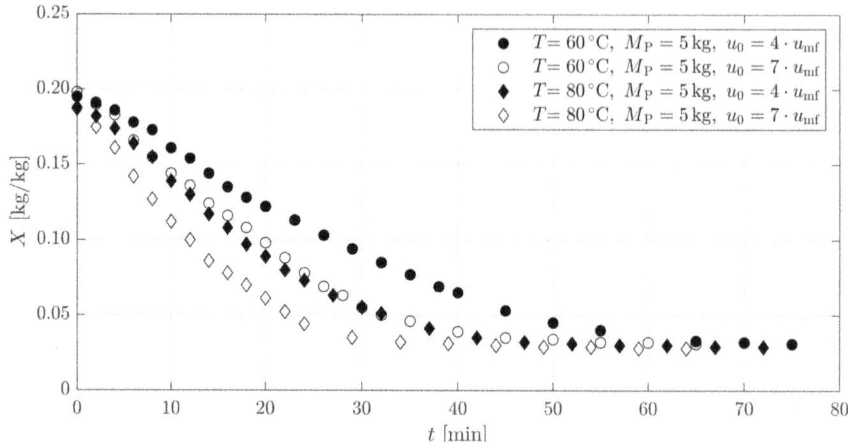


Figure B.1: Moisture content ($X$) of Cellets against drying time ($t$), showing influence of gas velocity ($u$) and drying temperature ($T$) on the drying kinetics in the *GF3* dryer.

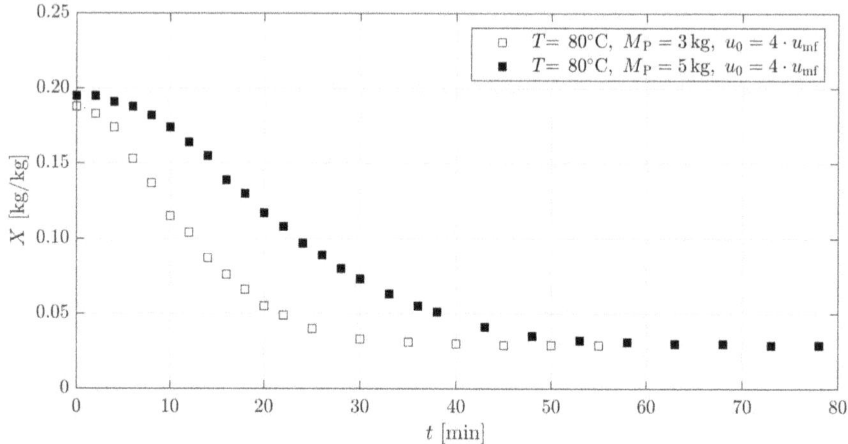


Figure B.2: Moisture content ($X$) of Cellets against drying time ($t$), showing influence of bed mass ($M_P$) on drying kinetics in the *GF3* dryer.

# Appendix C

# Definition of Transfer Coefficients and Transfer Areas

## Transfer Area - Particles and Suspension

Under the assumption of spherical particles, the transfer area between particles and suspension is the total surface area of all particles (Burgschweiger, 2000):

$$A_{\mathrm{PS}} = \frac{6\, M_{\mathrm{P}}}{d_{\mathrm{P}}\, \rho_P}\,. \tag{C.1}$$

## Transfer Area - Particles and Wall

The transfer area between particles and walls is defined as the wall area that is in contact with particles during operation of the fluidized bed dryer. It is calculated according to the geometry of the dryer:

$$A_{\mathrm{PW}} = \begin{cases} \pi\, d_{\mathrm{fb}}\, H_{\mathrm{fb}} & \text{for circular cross section} \\ (2w + 2b)\, H_{\mathrm{fb}} & \text{for rectangular cross section.} \end{cases} \tag{C.2}$$

Reference to symbols is given in Figure C.1.

## Mass Transfer - Suspension and Bubbles

Mass transfer coefficient and transfer surface area between suspension and bubbles ($\beta_{\mathrm{SB}} A_{\mathrm{SB}}$) are calculated according to Groenewold and Tsotsas (1997), using the theoretical number of transfer unit (NTU).

$$NTU_{\mathrm{SB}} = \frac{H_{\mathrm{fb}}}{H_{\mathrm{NTU=1}}} = \frac{\rho_G}{\dot{M}_{\mathrm{G}}} \beta_{\mathrm{SB}} A_{\mathrm{SB}}\,, \tag{C.3}$$

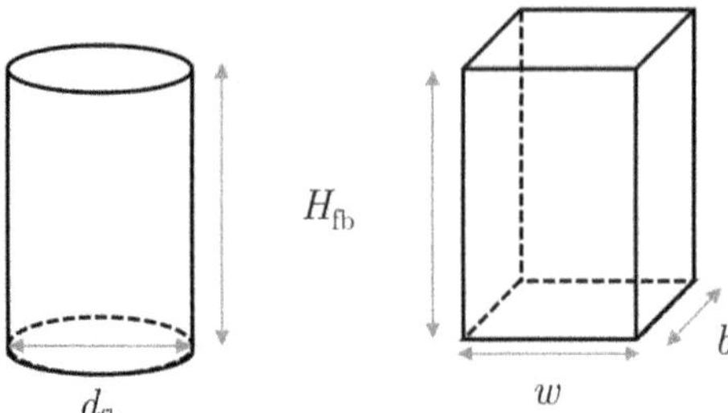


Figure C.1: Schematic shapes and symbols of dimensions of implemented dryer geometries.

where the height above the distributor at which $NTU_{\text{SB}}$ equals 1 is denoted by $H_{\text{NTU}=1}$. Groenewold and Tsotsas (1997) defined $H_{\text{NTU}=1} = 0.05\,\text{m}$, representing the height of the outflow hole of the continuous fluidized bed dryer they used. This assumption was successfully applied by Alaathar (2017), Burgschweiger (2000), Chen et al. (2017) for very similar particles and dryer geometries. Sensitivity analysis with regard to the influence $H_{\text{NTU}=1}$ is discussed in chapter 5.2.5.

## Mass Transfer - Particles and Suspension

The mass transfer coefficient between particles and suspension is calculated according to Gnielinski (1980):

$$\beta_{\text{PS}} = \frac{Sh'_{\text{PS}} D}{d_{\text{P}}}, \tag{C.4}$$

with $D$ representing the binary diffusion coefficient of water in air (see equation 2.37) and:

$$Re = \frac{Re_{\text{mf}}}{\varepsilon_{\text{mf}}}, \tag{C.5}$$

$$Sc = \frac{\nu_{\text{G}}}{D}, \tag{C.6}$$

$$Sh_{\text{lam}} = 0.664 \cdot \sqrt[3]{Sc}\sqrt{Re_{\text{t}}}, \tag{C.7}$$

$$Sh_{\text{turb}} = \frac{0.037 \cdot Re^{0.8} \cdot Sc}{1 + 2.443 \cdot Re^{-0.1}\left(Sc^{2/3} - 1\right)}, \tag{C.8}$$

$$Sh_{\text{single particle}} = 2 + \sqrt{Sh_{\text{lam}}^2 + Sh_{\text{turb}}^2}, \tag{C.9}$$

$$Sh_{\text{PS}} = \left(1 + 1.5\left(1 - \varepsilon_{\text{mf}}\right)\right) Sh_{\text{single particle}}. \tag{C.10}$$

Back mixing of particles is accounted for by the apparent Sherwood number according to Groenewold and Tsotsas (1997) and Burgschweiger (2000).

$$Sh'_{\mathrm{PS}} = \frac{Re_{\mathrm{op}}\, Sc}{A/F} \ln\left(1 + \frac{Sh_{\mathrm{PS}}\, A/F}{Re_{\mathrm{op}} Sc}\right), \tag{C.11}$$

$$Re_{\mathrm{op}} = \frac{u\, d_{\mathrm{P}}}{\nu_{\mathrm{G}}}, \tag{C.12}$$

$$A/F = \frac{A_{\mathrm{PS}}}{A_{\mathrm{fb}}} = 6\,(1-\varepsilon)\,\frac{H_{\mathrm{fb}}}{d_{\mathrm{P}}}. \tag{C.13}$$

## Heat Transfer - Particles and Suspension

The heat transfer coefficient between particles and suspension is calculated according to Gnielinski (1980).

$$\alpha_{\mathrm{PS}} = \frac{Nu'_{\mathrm{PS}}\, \lambda_{\mathrm{G}}}{d_{\mathrm{P}}}, \tag{C.14}$$

with:

$$Re = \frac{Re_{\mathrm{mf}}}{\varepsilon_{\mathrm{mf}}}, \tag{C.15}$$

$$Pr = \frac{\nu_{\mathrm{G}}\, c_{p,\mathrm{G}}\, \rho_G}{\lambda_{\mathrm{G}}}, \tag{C.16}$$

$$Nu_{\mathrm{lam}} = 0.664 \cdot \sqrt[3]{Sc}\sqrt{Re_{\mathrm{t}}}, \tag{C.17}$$

$$Nu_{\mathrm{turb}} = \frac{0.037 \cdot Re^{0.8} \cdot Pr}{1 + 2.443 \cdot Re^{-0.1}\left(Pr^{2/3} - 1\right)}, \tag{C.18}$$

$$Nu_{\mathrm{single\ particle}} = 2 + \sqrt{Nu_{\mathrm{lam}}^2 + Nu_{\mathrm{turb}}^2}, \tag{C.19}$$

$$Nu_{\mathrm{PS}} = (1 + 1.5\,(1 - \varepsilon_{\mathrm{mf}}))\, Nu_{\mathrm{single\ particle}}. \tag{C.20}$$

Back mixing of particles is accounted for by the apparent Nusselt number according to Groenewold and Tsotsas (1997) and Burgschweiger (2000):

$$Nu'_{\mathrm{PS}} = \frac{Re_{\mathrm{op}}\, Pr}{A/F} \ln\left(1 + \frac{Nu_{\mathrm{PS}}\, A/F}{Re_{\mathrm{op}} Pr}\right), \tag{C.21}$$

$$Re_{\mathrm{op}} = \frac{u\, d_{\mathrm{P}}}{\nu_{\mathrm{G}}}, \tag{C.22}$$

$$A/F = \frac{A_{\mathrm{PS}}}{A_{\mathrm{fb}}} = 6\,(1-\varepsilon)\,\frac{H_{\mathrm{fb}}}{d_{\mathrm{P}}}. \tag{C.23}$$

## Heat Transfer - Suspension and Bubbles

The heat transfer coefficient and transfer area between suspension and bubble phase is calculated using equation C.3 and the analogy between heat and mass transfer (Burgschweiger,

2000, Groenewold and Tsotsas, 1997):

$$\alpha_{\mathrm{SB}} A_{\mathrm{SB}} = \rho_G \cdot c_{p,\mathrm{G}} \cdot \beta_{\mathrm{SB}} A_{\mathrm{SB}} \cdot Le^m \,, \tag{C.24}$$

with:

$$Le = \frac{\lambda_{\mathrm{G}}}{c_{p,\mathrm{G}}\, \rho_G\, D} \,, \tag{C.25}$$

$$m = 1/3 \,. \tag{C.26}$$

## Heat Transfer - Particle and Wall

The heat transfer coefficient between particles and wall is calculated according to Martin (2010):

$$\alpha_{\mathrm{PW}} = \frac{Nu_{\mathrm{PW}} \cdot \lambda_{\mathrm{G}}}{d_{\mathrm{P}}} \,, \tag{C.27}$$

with:

$$Nu_{\mathrm{PW}} = (1-\varepsilon)\, Z\, \left(1 - e^{-N}\right) \,, \tag{C.28}$$

$$Z = \frac{1}{6} \frac{\rho_P \cdot c_{p,\mathrm{P,wet}}}{\lambda_{\mathrm{G}}} \sqrt{\frac{g\, d_{\mathrm{P}}^3\, (\varepsilon - \varepsilon_{\mathrm{mf}})}{5\,(1-\varepsilon_{\mathrm{mf}})\,(1-\varepsilon)}} \,, \tag{C.29}$$

$$N = \frac{Nu_{\mathrm{PW,max}}}{C_K \cdot Z} \tag{C.30}$$

$$C_K = 2.6 \tag{C.31}$$

$$Nu_{\mathrm{PW,max}} = 4\left(\left(1 + \frac{2l}{d_{\mathrm{P}}}\right) \ln\left(1 + \frac{d_{\mathrm{P}}}{2l}\right)\right) - 1 \,, \tag{C.32}$$

$$l = 2\left(\frac{2}{\gamma} - 1\right) \sqrt{\frac{2\pi\, R\, T[\mathrm{K}]}{\tilde{M}_{\mathrm{G}}}} \frac{\lambda_{\mathrm{G}}}{p[\mathrm{Pa}] \cdot \left(2 c_{p,\mathrm{G}} - R/\tilde{M}_{\mathrm{G}}\right)} \,, \tag{C.33}$$

$$\gamma = \left(1 + 10^{0.6 - \frac{1000\,[\mathrm{K}]}{T[\mathrm{K}]} C_{\mathrm{A}}^{-1}}\right)^{-1} \,, \tag{C.34}$$

$$C_{\mathrm{A}} = 2.8 \text{ (for air)} \,. \tag{C.35}$$

# Appendix D

# Supplementary Simulation Data

Detailed results of all validation cases, discussed in chapter 5.1, are summarized in the following tables.
Subsequently, sensitivity analysis data, not shown in chapter 5.2, are illustrated in following figures. All figures are structured equally. The topmost plot shows data for the reference case with $\gamma$-$Al_2O_3$ particles in the *GF3* lab-scale dryer (reference case GF3_A_02). Reference case GF3_C_03 is located in the middle and case VFB_F_02 represented in the bottom plot. The left vertical axis shows the particle moisture content at the dryer outlet ($X_P$), the right vertical axis represents the particle temperature at the outlet ($T_P$).

Table D.1: Overview of experimental and simulation results including relative deviations of validation experiments and simulation of cases in *GF3* lab-scale dryer; negative relative deviations indicate over predictions of the model, positive relative deviations indicate model under predictions.

| Exp. name | $H_{\mathrm{fb,exp}}$ [cm] | $H_{\mathrm{fb,sim}}$ [cm] | $\Delta H_{\mathrm{fb}}$ [%] | $T_{\mathrm{P,exp}}$ [K] | $T_{\mathrm{P,sim}}$ [K] | $\Delta T_{\mathrm{P}}$ [%] | $X_{\mathrm{P,exp}}$ [kg kg$^{-1}$] | $X_{\mathrm{P,sim}}$ [kg kg$^{-1}$] | $\Delta X_{\mathrm{P}}$ [%] |
|---|---|---|---|---|---|---|---|---|---|
| GF3_A_01 | 13.0 | 12.4 | 4.47 | 295.60 | 294.95 | 0.22 | 0.170 | 0.144 | 15.16 |
| GF3_A_02 | 13.0 | 12.4 | 4.35 | 301.55 | 301.37 | 0.06 | 0.291 | 0.281 | 3.51 |
| GF3_A_03 | 13.0 | 12.4 | 4.27 | 308.68 | 314.67 | -1.94 | 0.141 | 0.139 | 1.45 |
| GF3_A_04 | 13.0 | 12.4 | 4.32 | 308.02 | 312.53 | -1.47 | 0.172 | 0.182 | -5.68 |
| GF3_A_05 | 13.0 | 12.4 | 4.28 | 308.98 | 313.53 | -1.47 | 0.150 | 0.150 | 0.03 |
| GF3_A_06 | 8.0 | 7.7 | 4.19 | 307.33 | 318.81 | -3.74 | 0.138 | 0.132 | 4.38 |
| GF3_C_01 | 8.1 | 8.2 | -2.20 | 301.62 | 303.56 | -0.64 | 0.087 | 0.060 | 31.23 |
| GF3_C_02 | 8.1 | 8.2 | -2.24 | 301.25 | 302.40 | -0.38 | 0.073 | 0.069 | 5.19 |
| GF3_C_03 | 7.2 | 7.4 | -3.24 | 306.52 | 311.91 | -1.76 | 0.081 | 0.064 | 21.04 |
| GF3_C_04 | 7.2 | 7.4 | -3.24 | 302.91 | 306.55 | -1.20 | 0.124 | 0.092 | 25.45 |
| GF3_C_05 | 8.1 | 8.3 | -3.14 | 315.69 | 318.16 | -0.78 | 0.061 | 0.054 | 10.87 |
| GF3_C_06 | 8.1 | 8.3 | -3.05 | 311.69 | 312.85 | -0.37 | 0.068 | 0.070 | -3.75 |
| GF3_C_07 | 7.2 | 7.4 | -3.84 | 325.43 | 334.32 | -2.73 | 0.046 | 0.046 | -1.05 |
| GF3_C_08 | 8.1 | 8.4 | -3.88 | 336.58 | 336.53 | 0.02 | 0.044 | 0.052 | -18.06 |
| GF3_C_09 | 8.1 | 8.4 | -3.83 | 326.78 | 330.95 | -1.28 | 0.062 | 0.056 | -8.70 |

Table D.2: Overview of experimental and simulation results including relative deviations of validation experiments and simulations of cases in VFB pilot-plant-scale dryer; negative relative deviations indicate over predictions of the model, positive relative deviations indicate model under predictions.

| Exp. name | $H_{\text{fb,exp}}$ [cm] | $H_{\text{fb,sim}}$ [cm] | $\Delta H_{\text{fb}}$ [%] | $T_{\text{P,exp}}$ [K] | $T_{\text{P,sim}}$ [K] | $\Delta T_{\text{P}}$ [%] | $X_{\text{P,exp}}$ [kg kg$^{-1}$] | $X_{\text{P,sim}}$ [kg kg$^{-1}$] | $\Delta X_{\text{P}}$ [%] |
|---|---|---|---|---|---|---|---|---|---|
| VFB_C_01 | 6.8 | 6.67 | 1.88 | 296.79 | 296.50 | 0.10 | 0.104 | 0.109 | -4.68 |
| VFB_C_02 | 7.2 | 6.29 | 12.6 | 294.91 | 295.01 | -0.06 | 0.157 | 0.169 | -7.10 |
| VFB_C_03 | 6.9 | 6.88 | 0.32 | 305.88 | 317.23 | -3.71 | 0.059 | 0.057 | 3.51 |
| VFB_C_04 | 6.9 | 6.96 | -0.93 | 307.60 | 318.60 | -3.58 | 0.056 | 0.051 | 8.60 |
| VFB_C_05 | 6.9 | 6.94 | -0.56 | 303.90 | 309.10 | -1.71 | 0.081 | 0.075 | 6.75 |
| VFB_C_06 | 7.0 | 7.18 | -2.63 | 310.64 | 315.73 | -1.64 | 0.054 | 0.063 | -15.91 |
| VFB_C_07 | 7.0 | 7.20 | -2.80 | 298.05 | 298.35 | -0.10 | 0.111 | 0.096 | 13.61 |
| VFB_C_08 | 7.0 | 7.18 | -2.55 | 303.72 | 306.70 | -0.98 | 0.105 | 0.091 | 13.24 |
| VFB_C_09 | 7.0 | 7.19 | -2.71 | 309.55 | 318.34 | -2.84 | 0.063 | 0.055 | 12.04 |
| VFB_F_01 | 7.0 | 8.13 | -16.10 | 291.64 | 292.13 | -0.46 | 0.167 | 0.181 | -8.07 |
| VFB_F_02 | 7.0 | 7.16 | -2.21 | 291.78 | 292.17 | -0.39 | 0.186 | 0.176 | 5.71 |
| VFB_F_03 | 15.0 | 14.24 | 5.10 | 293.09 | 293.37 | -0.27 | 0.144 | 0.167 | -15.53 |
| VFB_F_04 | 15.0 | 16.26 | -8.42 | 291.60 | 292.46 | -0.97 | 0.163 | 0.176 | -8.16 |
| VFB_F_05 | 15.0 | 14.79 | 1.41 | 291.76 | 292.68 | -0.91 | 0.180 | 0.170 | 5.92 |
| VFB_F_06 | 15.0 | 15.63 | -4.22 | 294.48 | 294.38 | 0.11 | 0.037 | 0.031 | 16.11 |
| VFB_F_07 | 15.0 | 12.98 | 13.47 | 294.16 | 294.31 | -0.16 | 0.057 | 0.066 | -16.12 |

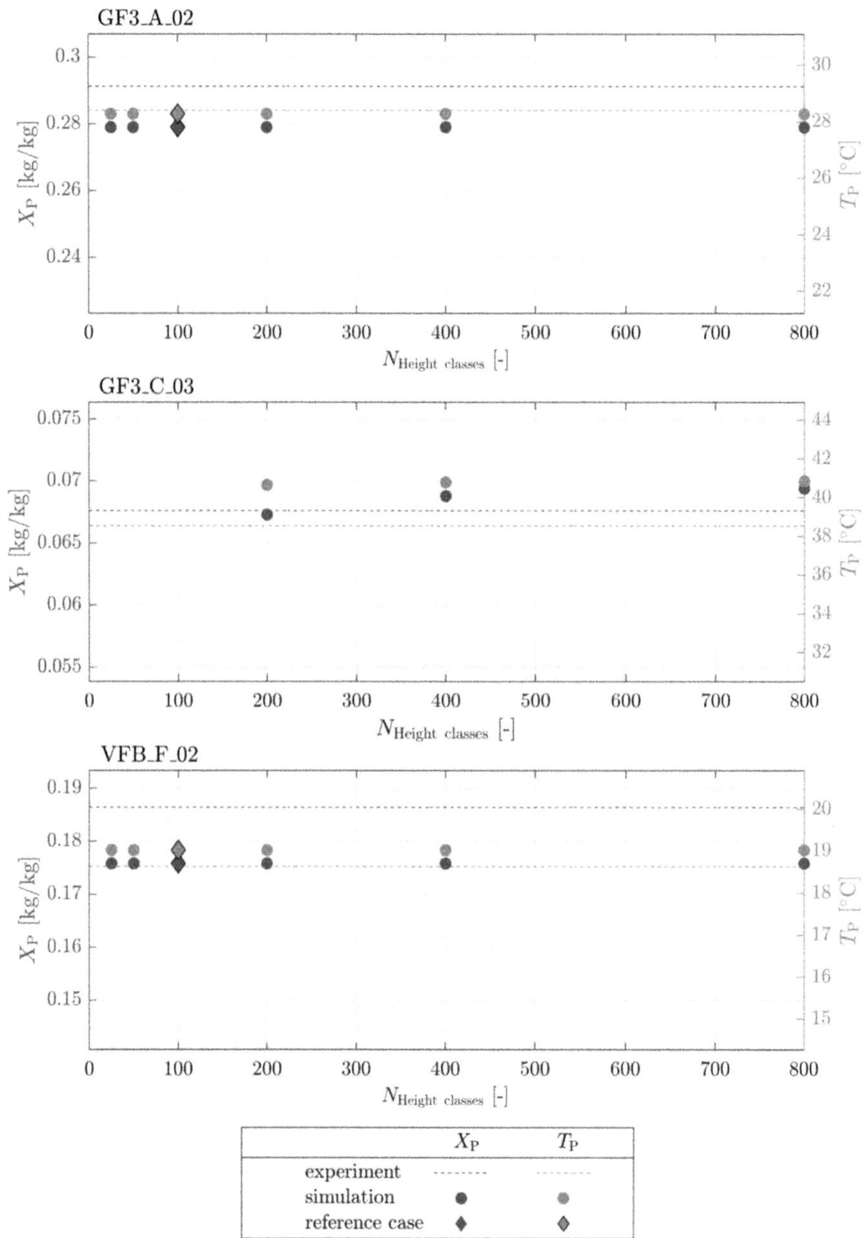


Figure D.1: Sensitivity analysis with regard to selected number of height classes.

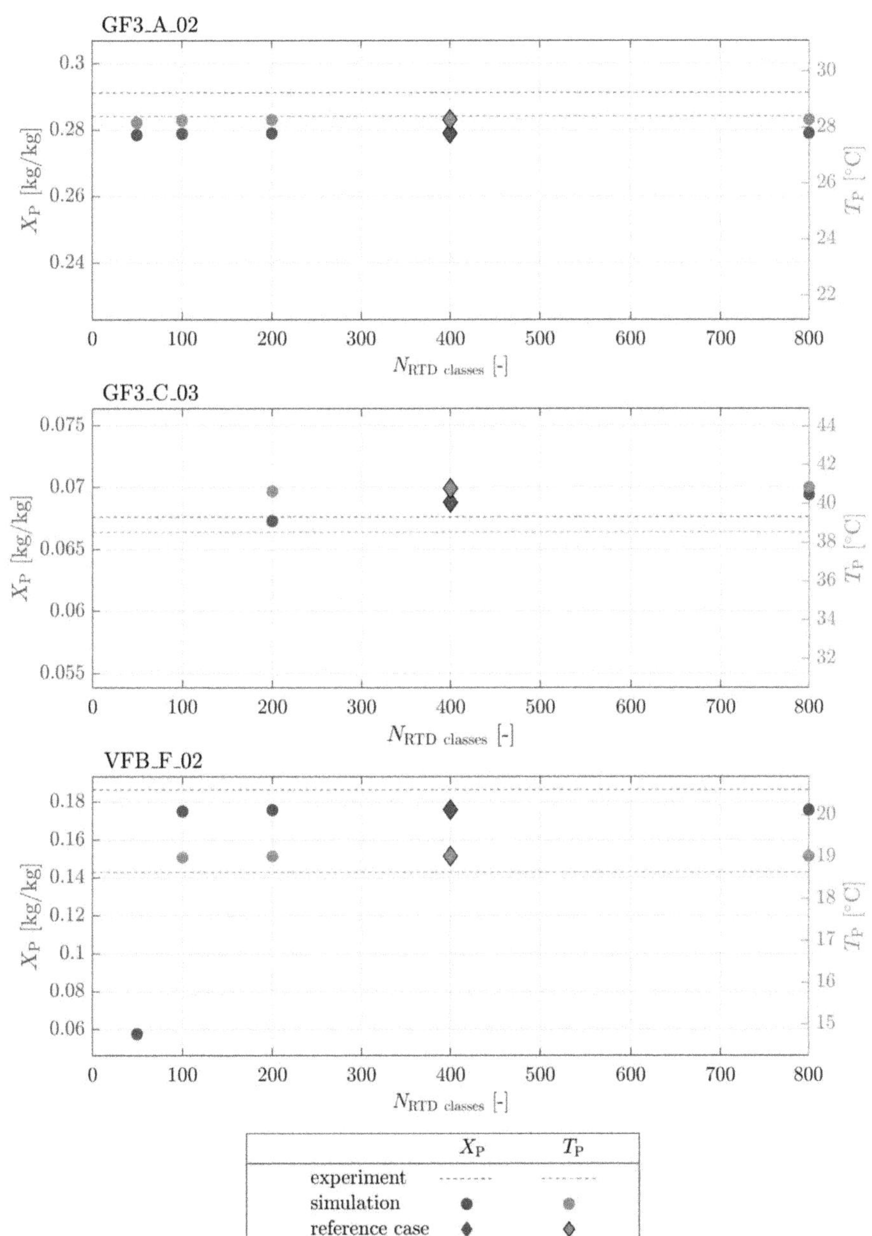


Figure D.2: Sensitivity analysis with regard to selected number of residence time distribution (RTD) classes.

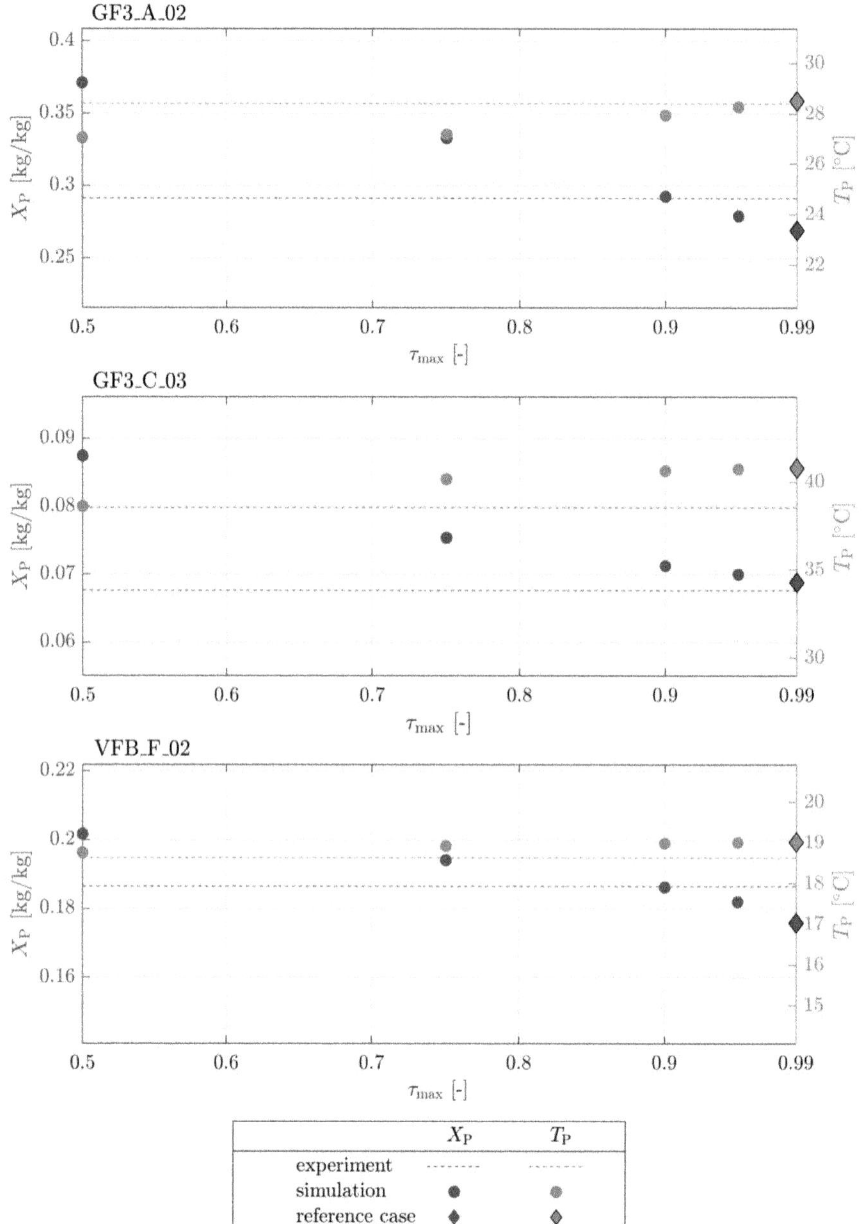


Figure D.3: Sensitivity analysis with regard to selected maximum of considered residence time ($\tau_{max}$).

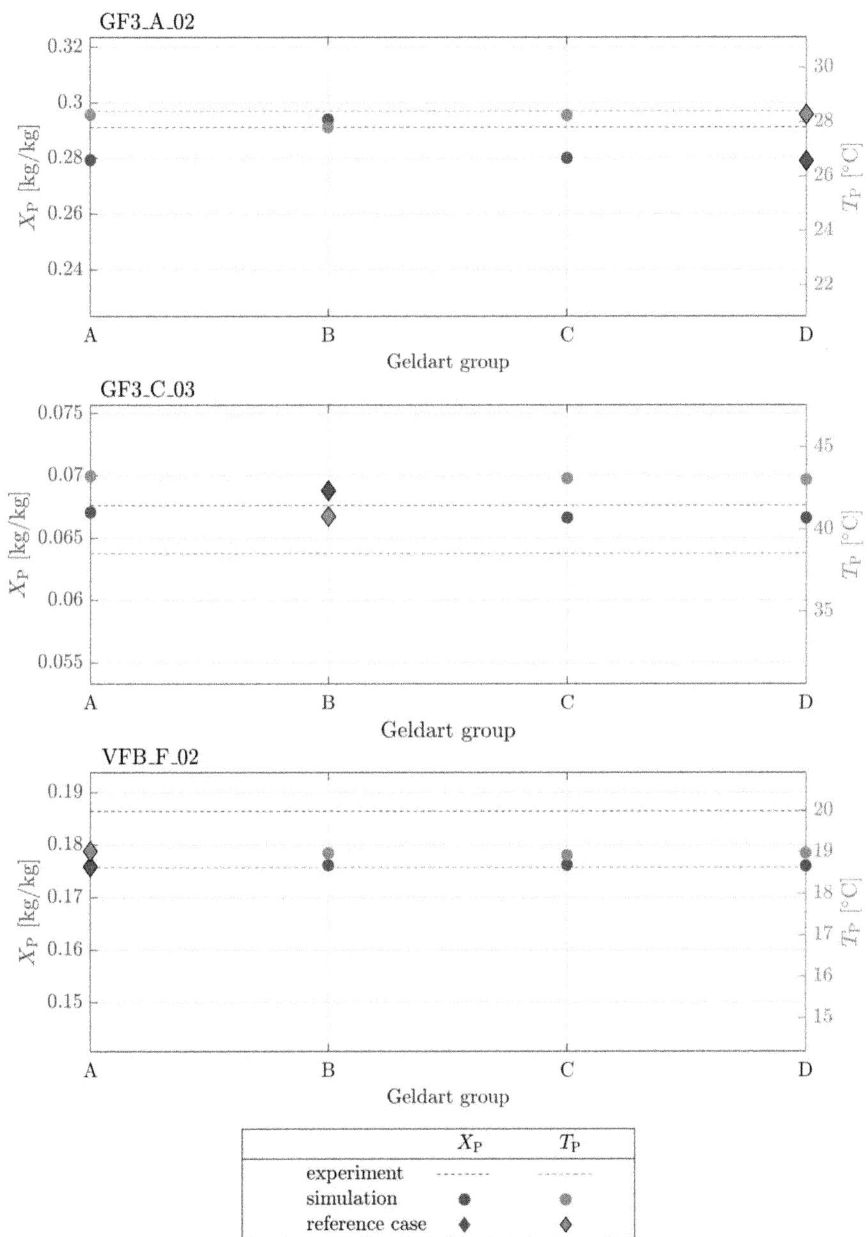


Figure D.4: Sensitivity analysis with regard to selected Geldart group.

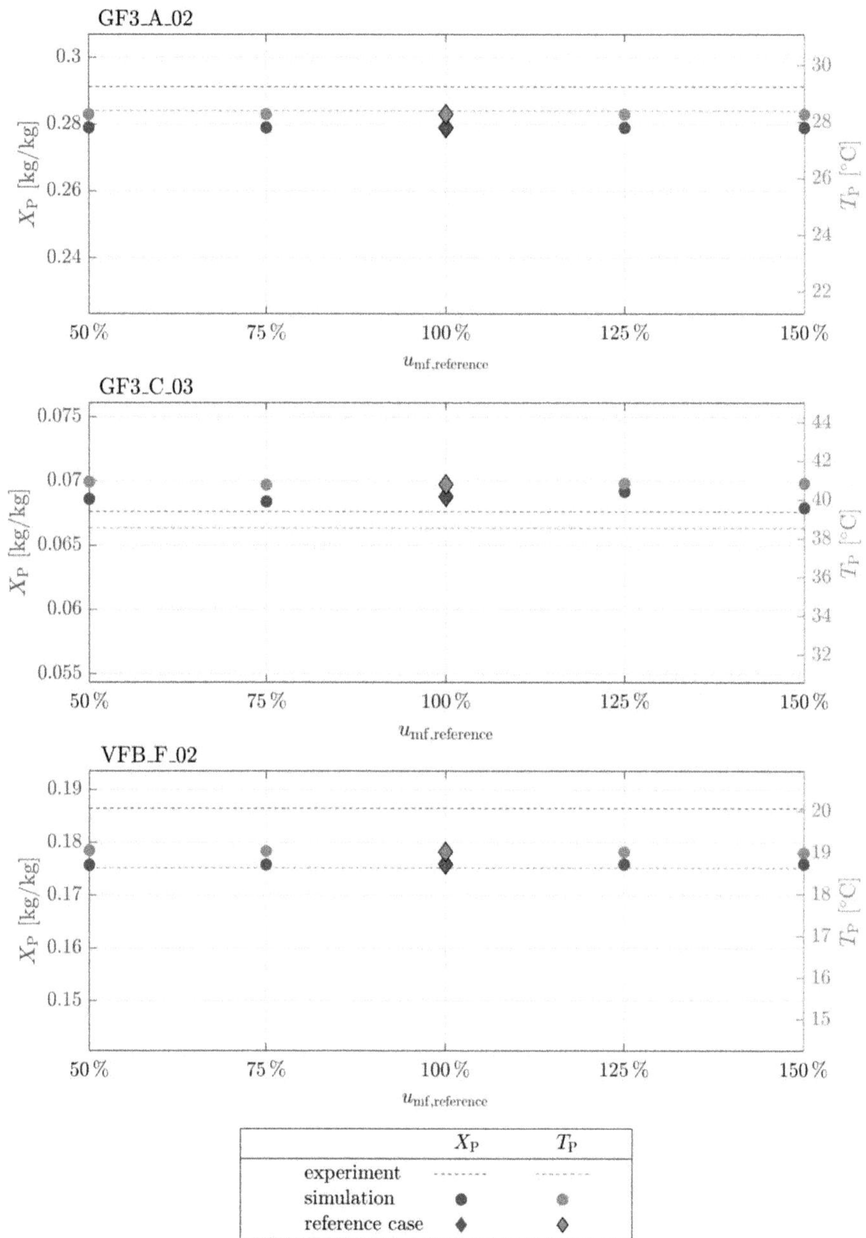


Figure D.5: Sensitivity analysis with regard to selected minimum fluidization velocity ($u_{mf}$).

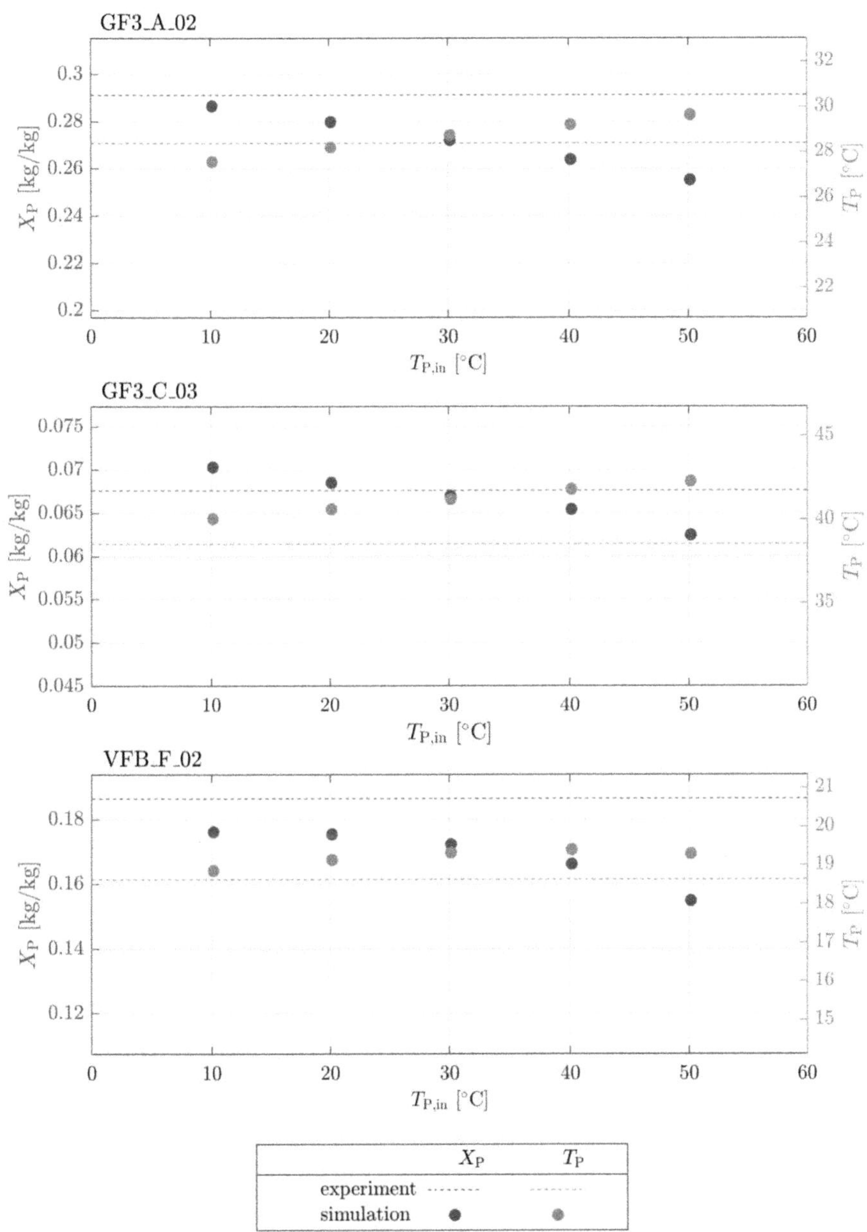


Figure D.6: Sensitivity analysis with regard to selected particle inlet temperature ($T_{P,in}$).

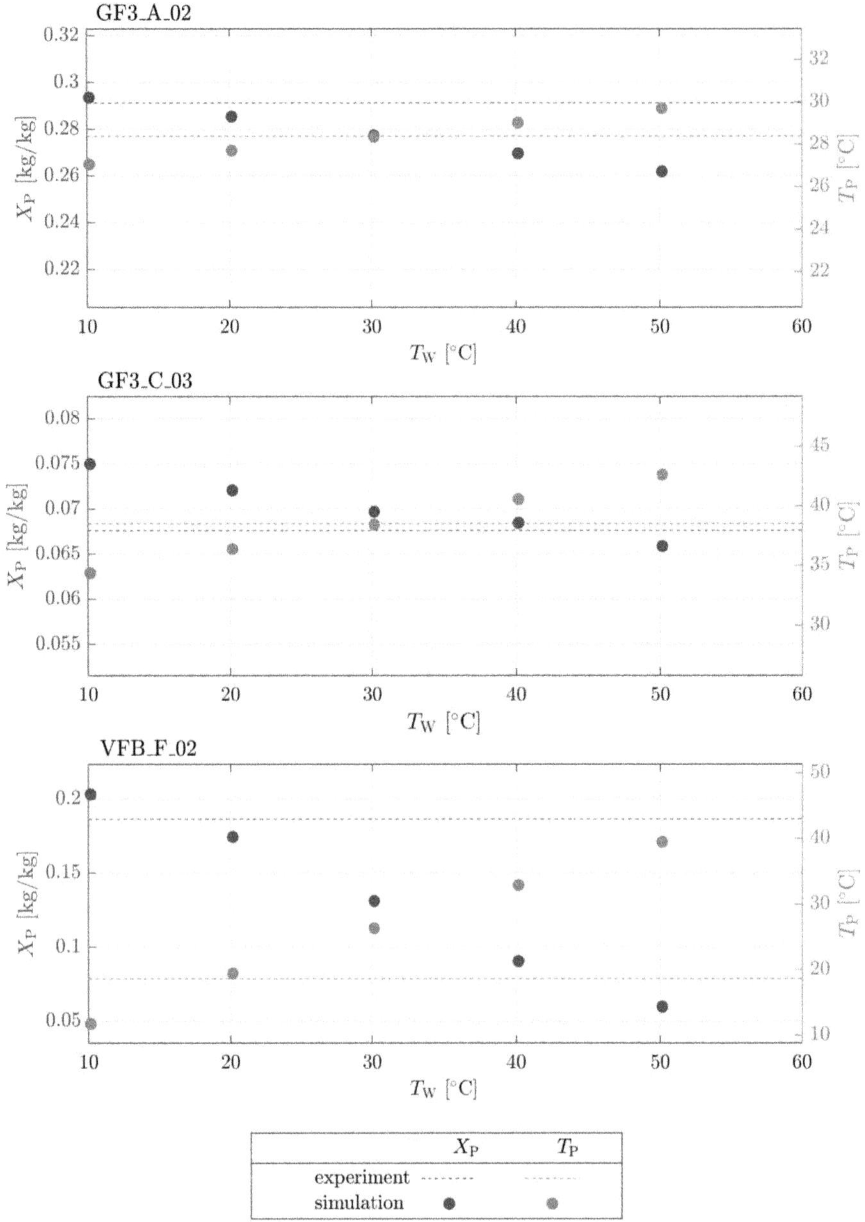


Figure D.7: Sensitivity analysis with regard to selected wall temperature ($T_W$).

# Bibliography


Abbasi, M. R., Shamiri, A., Hussain, M. A., and Kaboli, S. H. A. (2019). Dynamic process modeling and hybrid intelligent control of ethylene copolymerization in gas phase catalytic fluidized bed reactors. *Journal of Chemical Technology & Biotechnology*, 94(8):2433–2451.

Agu, C. E., Tokheim, L.-A., Eikeland, M., and Moldestad, B. M. (2019). Improved models for predicting bubble velocity, bubble frequency and bed expansion in a bubbling fluidized bed. *Chemical Engineering Research and Design*, 141:361–371.

Alaathar, I. (2017). *Fließschema-Simulation der kontinuierlichen Wirbelschichttrocknung mit verteilten Parametern*, volume v.9 of *SPE-Schriftenreihe*. Cuvillier Verlag, Göttingen, 1st edition.

Alaathar, I., Hartge, E.-U., Heinrich, S., and Werther, J. (2013). Modeling and flowsheet simulation of continuous fluidized bed dryers. *Powder Technology*, 238:132–141.

Alaathar, I., Heinrich, S., and Hartge, E.-U. (2020). Continuous Fluidized Bed Drying: Advanced Modeling and Experimental Investigations. In Nagy, Z. K., El Hagrasy, A., and Litster, J., editors, *Continuous Pharmaceutical Processing*, volume 42 of *AAPS Advances in the Pharmaceutical Sciences Series*, pages 301–359. Springer International Publishing, Cham.

AlOthman, Z. (2012). A Review: Fundamental Aspects of Silicate Mesoporous Materials. *Materials*, 5(12):2874–2902.

Anderson, D. G. (1965). Iterative Procedures for Nonlinear Integral Equations. *Journal of the ACM (JACM)*, 12(4):547–560.

Attari Moghaddam, A., Kharaghani, A., Tsotsas, E., and Prat, M. (2018). A pore network study of evaporation from the surface of a drying non-hygroscopic porous medium. *AIChE Journal*, 64(4):1435–1447.

Bachmann, P., Bück, A., and Tsotsas, E. (2016). Investigation of the residence time behavior of particulate products and correlation for the Bodenstein number in horizontal fluidized beds. *Powder Technology*, 301:1067–1076.

Bachmann, P., Bück, A., and Tsotsas, E. (2017). Experimental investigation and correlation of the Bodenstein number in horizontal fluidized beds with internal baffles. *Powder Technology*, 308:378–387.

Bakshi, A., Altantzis, C., Glicksman, L. R., and Ghoniem, A. F. (2017). Gas-flow distribution in bubbling fluidized beds: CFD-based analysis and impact of operating conditions. *Powder Technology*, 316:500–511.

Bargieł, M. and Tory, E. M. (2013). Extension of the Richardson–Zaki equation to suspensions of multisized irregular particles. *International Journal of Mineral Processing*, 120:22–25.

Barletta, D. and Poletto, M. (2012). Aggregation phenomena in fluidization of cohesive powders assisted by mechanical vibrations. *Powder Technology*, 225:93–100.

Barletta, D., Russo, P., and Poletto, M. (2013). Dynamic response of a vibrated fluidized bed of fine and cohesive powders. *Powder Technology*, 237:276–285.

Bi, H. T. (2007). A critical review of the complex pressure fluctuation phenomenon in gas–solids fluidized beds. *Chemical Engineering Science*, 62(13):3473–3493.

Bi, H. T., Ellis, N., Abba, I. A., and Grace, J. R. (2000). A state-of-the-art review of gas–solid turbulent fluidization. *Chemical Engineering Science*, 55(21):4789–4825.

Brennan, W., Jacobson, M., Book, G., Briens, C., and Briens, L. (2008). Development of a tribolelectric procedure for the measurement of mixing and drying in a vibrated fluidized bed. *Powder Technology*, 181(2):178–185.

Brod, F., Park, K. J., and de Almeida, R. G. (2004). Image Analysis to Obtain the Vibration Amplitude and the Residence Time Distribution of a Vibro-Fluidized Dryer. *Food and Bioproducts Processing*, 82(2):157–163.

Brunauer, S., Emmett, P. H., and Teller, E. (1938). Adsorption of Gases in Multimolecular Layers. *Journal of the American Chemical Society*, 60(2):309–319.

Burgschweiger, J. (2000). *Modellierung des statischen und dynamischen Verhaltens von kontinuierlich betriebenen Wirbelschichttrocknern: Zugl.: Magdeburg, Univ., Fak. für Verfahrens- und Systemtechnik, Diss., 2000*, volume 665 of *Fortschritt-Berichte / VDI Reihe 3, Verfahrenstechnik*. VDI-Verl., Düsseldorf.

Burgschweiger, J., Groenewold, H., Hirschmann, C., and Tsotsas, E. (1999). From hygroscopic single particle to batch fluidized bed drying kinetics. *The Canadian Journal of Chemical Engineering*, 77(2):333–341.

Burgschweiger, J. and Tsotsas, E. (2002). Experimental investigation and modelling of continuous fluidized bed drying under steady-state and dynamic conditions. *Chemical Engineering Science*, 57(24):5021–5038.

Cano-Pleite, E., Hernández-Jiménez, F., de Vega, M., and Acosta-Iborra, A. (2014). Experimental study on the motion of isolated bubbles in a vertically vibrated fluidized bed. *Chemical Engineering Journal*, 255:114–125.

Cano-Pleite, E., Shimizu, Y., Acosta-Iborra, A., and Mawatari, Y. (2016). Effect of vertical vibration and particle size on the solids hold-up and mean bubble behavior in a pseudo-2D fluidized bed. *Chemical Engineering Journal*, 304:384–398.

Chen, H., Rustagi, S., Diep, E., Langrish, T. A., and Glasser, B. J. (2018). Scale-up of fluidized bed drying: Impact of process and design parameters. *Powder Technology*, 339:8–16.

Chen, K., Bachmann, P., Bück, A., Jacob, M., and Tsotsas, E. (2017). Experimental study and modeling of particle drying in a continuously-operated horizontal fluidized bed. *Particuology*, 34:134–146.

Chen, X. D. (2008). The Basics of a Reaction Engineering Approach to Modeling Air-Drying of Small Droplets or Thin-Layer Materials. *Drying Technology*, 26(6):627–639.

Chen, X. D. and Lin, S. X. Q. (2004). The Reaction Engineering Approach to Modelling Drying of Milk Droplets. In *Proceedings of the 14th International Drying Symposium (IDS 2004)*, pages 1644–1651.

Chen, X. D. and Lin, S. X. Q. (2005). Air drying of milk droplet under constant and time-dependent conditions. *AIChE Journal*, 51(6):1790–1799.

Chen, X. D. and Putranto, A. (2013). *Modelling drying processes: A reaction engineering approach.* Cambridge University Press, Cambridge.

Chen, X. D. and Xie, G. Z. (1997). Fingerprints of the Drying Behaviour of Particulate or Thin Layer Food Materials Established Using a Reaction Engineering Model. *Food and Bioproducts Processing*, 75(4):213–222.

Chew, J. H., Fu, N., Woo, M. W., Patel, K., Selomulya, C., and Chen, X. D. (2013). Capturing the effect of initial concentrations on the drying kinetics of high solids milk using reaction engineering approach. *Dairy Science and Technology*, 93(4-5):415–430.

Clift, R., Ghadiri, M., Monteiro, J. L., Tan, B., and Thambimuthu, K. V. (1983). Interphase Transfer and Interstitial Flow in Bubbling Beds, inferred from Measurements in Aerosol Capture. *Proceeding of 4th International Fluidization Conference, Koshikojima, Japan*, pages 77–85.

Cruz, M. A. A., Passos, M. L., and Ferreira, W. R. (2005). Final Drying of Whole Milk Powder in Vibrated-Fluidized Beds. *Drying Technology*, 23(9-11):2021–2037.

Daleffe, R. V., Ferreira, M. C., and Freire, J. T. (2005). Drying of Pastes in Vibro-Fluidized Beds: Effects of the Amplitude and Frequency of Vibration. *Drying Technology*, 23(9-11):1765–1781.

Danckwerts, P. V. (1953). Continuous flow systems: Distribution of residence time. *Chemical Engineering Science*, 2(1):1–13.

Davidson, J. F. and Harrison, D. (1966). The behaviour of a continuously bubbling fluidised bed. *Chemical Engineering Science*, 21(9):731–738.

Davis, P. J. (1959). Leonhard Euler's Integral: A Historical Profile of the Gamma Function: In Memoriam: Milton Abramowitz. *The American Mathematical Monthly*, 66(10):849.

Defraeye, T., Blocken, B., Derome, D., Nicolai, B., and Carmeliet, J. (2012). Convective heat and mass transfer modelling at air–porous material interfaces: Overview of existing methods and relevance. *Chemical Engineering Science*, 74:49–58.

Diez, E., Kieckhefen, P., Meyer, K., Bück, A., Tsotsas, E., and Heinrich, S. (2019). Particle dynamics in a multi-staged fluidized bed: Particle transport behavior on micro-scale by discrete particle modelling. *Advanced Powder Technology*, 30(10):2014–2031.

DIN (2001). DIN 51705:2001-06: Testing of solid fuels - Determination of the bulk density of solid fuels.

DIN (2003). DIN EN ISO 5167-2:2003: Durchflussmessung von Fluiden mit Drosselgeräten in volldurchströmten Leitungen mit Kreisquerschnitt - Teil 2: Blenden.

DIN (2012). DIN ISO 9276-6:2008: Darstellung der Ergebnisse von Partikelgrößenanalysen –Teil 6: Deskriptive und quantitative Darstellung der Form und Morphologie von Partikeln.

DIN (2019a). DIN 66137-1:2019-03: Bestimmung der Dichte fester Stoffe - Teil 1: Grundlagen.

DIN (2019b). DIN 66137-2:2019-03: Bestimmung der Dichte fester Stoffe - Teil 2: Gaspyknometrie.

Dosta, M., Litster, J. D., and Heinrich, S. (2020). Flowsheet simulation of solids processes: Current status and future trends. *Advanced Powder Technology*, 31(3):947–953.

Dry, R. J., Judd, M. R., and Shingles, T. (1984). Bubble Velocities in Fluidized Beds of Fine, Dense Powders. *Powder Technology*, 39(1):69–75.

Eccles, E. R. A. and Mujumdar, A. S. (1992). Cylinder to bed heat transfer in aerated vibrated beds of small particles. *Drying Technology*, 10(1):139–164.

Eccles, E. R. A. and Mujumdar, A. S. (1997). Bubble Phenomena in Aerated Vibrated Beds of Small Particles. *Drying Technology*, 15(1):95–110.

Ertekin, C. and Firat, M. Z. (2017). A comprehensive review of thin-layer drying models used in agricultural products. *Critical reviews in food science and nutrition*, 57(4):701–717.

Fabich, H. T., Sederman, A. J., and Holland, D. J. (2017). Study of bubble dynamics in gas-solid fluidized beds using ultrashort echo time (UTE) magnetic resonance imaging (MRI). *Chemical Engineering Science*, 172:476–486.

Felipe, C. A. S. and Rocha, S. C. S. (2004). Time series analysis of pressure fluctuation in gas-solid fluidized beds. *Brazilian Journal of Chemical Engineering*, 21(3):497–507.

Finzer, J., Sfredo, M. A., Sousa, G., and Limaverde, J. R. (2007). Dispersion coefficient of coffee berries in vibrated bed dryer. *Journal of Food Engineering*, 79(3):905–912.

Gao, Y., Muzzio, F. J., and Ierapetritou, M. G. (2012). A review of the Residence Time Distribution (RTD) applications in solid unit operations. *Powder Technology*, 228:416–423.

Geldart, D. (1973). Types of gas fluidization. *Powder Technology*, 7(5):285–292.

Gidaspow, D., Lu, H. L., and Mostofi, R. (2001). Large scale oscillations or gravity waves in risers and bubbling beds. In Li, J., Kwauk, M., and Yang, W. C., editors, *Fluidization X*, pages 317–324. Engineering Foundation, New York.

Gleiss, M., Hammerich, S., Kespe, M., and Nirschl, H. (2017). Application of the dynamic flow sheet simulation concept to the solid-liquid separation: Separation of stabilized slurries in continuous centrifuges. *Chemical Engineering Science*, 163:167–178.

Gnielinski, V. (1980). Wärme- und Stoffübertragung in Festbetten. *Chemie Ingenieur Technik*, 52(3):228–236.

Groenewold, H. and Tsotsas, E. (1997). A new model for fluid bed drying. *Drying Technology*, 15(6-8):1687–1698.

Groenewold, H. and Tsotsas, E. (1999). Predicting apparent Sherwood numbers in fluidized beds. *Drying Technology*, 17(7-8):1557–1570.

Groenewold, H. and Tsotsas, E. (2007). Drying in fluidized beds with immersed heating elements. *Chemical Engineering Science*, 62(1-2):481–502.

Gündoğdu, Ö. and Tüzün, U. (2006). Gas Fluidisation of Nano-particle Assemblies: Modified Geldart classification to account for multiple-scale fluidisation of agglomerates and clusters. *KONA Powder and Particle Journal*, 24(0):3–14.

Gupta, R. and Mujumdar, A. S. (1980). Aerodynamics of a vibrated fluid bed. *The Canadian Journal of Chemical Engineering*, 58(3):332–338.

Haghgoo, M. R., Bergstrom, D. J., and Spiteri, R. J. (2018). Effect of particle stress tensor in simulations of dense gas–particle flows in fluidized beds. *Particuology*, 38:31–43.

Han, W., Mai, B., and Gu, T. (1991). Residence Time Distribution and Drying Characteristics of a Continous Vibro-Fluidized Bed. *Drying Technology*, 9(1):159–181.

Hashimoto, A., Stenström, S., and Kameoka, T. (2003). Simulation of Convective Drying of Wet Porous Materials. *Drying Technology*, 21(8):1411–1431.

Haus, J., Hartge, E.-U., Heinrich, S., and Werther, J. (2018). Dynamic flowsheet simulation for chemical looping combustion of methane. *International Journal of Greenhouse Gas Control*, 72:26–37.

Heinrich, S., editor (2020). *Dynamic flowsheet simulation of solids processes*. SPRINGER NATURE, Switzerland.

Hilligardt, K. (1986). *Zur Strömungsmechanik von Grobkornwirbelschichten*. Dissertation, Hamburg University of Technology, Hamburg.

Hilligardt, K. and Werther, J. (1986). Local bubble gas hold-up and expansion of gas-solid fluidized beds. *German Chemical Engineering*, 9:215–221.

Hilligardt, K. and Werther, J. (1987). The influence of temperature and solids properties on the size and growth of bubbles in gas fluidized beds. *Chemical Engineering & Technology*, 10:272–280.

Hull, R. L. and Rosenberg, A. E. V. (1960). Radiochemical Tracing of Fluid Catalyst Flow. *Industrial & Engineering Chemistry*, 52(12):989–992.

Idakiev, V. and Mörl, L. (2013). Study of Residence Time of Disperse Materials in Continuously Operating Fluidized Bed Apparatus. *Journal of Chemical Technology and Metallurgy*, 48(5):451–456.

Jabbari, Y., Tsotsas, E., Kirsch, C., and Kharaghani, A. (2019). Determination of the moisture transport coefficient from pore network simulations of spontaneous imbibition in capillary porous media. *Chemical Engineering Science*, 207:600–610.

Jacob, M. (2010). *Experimentelle Untersuchung sowie Beiträge zur Modellierung von Prozessen in Wirbelschichtrinnen am Beispiel der Sprühgranulation*. Dissertation, Otto-von-Guericke-Universität Magdeburg, Magdeburg.

Jia, D., Cathary, O., Peng, J., Bi, X., Lim, C. J., Sokhansanj, S., Liu, Y., Wang, R., and Tsutsumi, A. (2015). Fluidization and drying of biomass particles in a vibrating fluidized bed with pulsed gas flow. *Fuel Processing Technology*, 138:471–482.

Jin, H., Zhang, J., and Zhang, B. (2007). The effect of vibration on bed voidage behaviors in fluidized beds with large particles. *Brazilian Journal of Chemical Engineering*, 24(3):389–397.

Kage, H., Oba, M., Ishimatsu, H., Hironao Ogura, and Matsuno, Y. (1999). The effects of frequency and amplitude on the powder coating of fluidizing particles in a vibro-fluidized bed. *Advanced Powder Technology*, 10(1):77–87.

Karimipour, S. and Pugsley, T. (2011). A critical evaluation of literature correlations for predicting bubble size and velocity in gas–solid fluidized beds. *Powder Technology*, 205(1-3):1–14.

Kemp, I. C. and Oakley, D. E. (2002). Modelling of particulate drying in theory and practice. *Drying Technology*, 20(9):1699–1750.

Khanali, M., Rafiee, S., Jafari, A., and Banisharif, A. (2012). Study of Residence Time Distribution of Rough Rice in a Plug Flow Fluid Bed Dryer. *International Journal of Advanced Science and Technology*, 48:103–114.

Kharaghani, A., Metzger, T., and Tsotsas, E. (2012). An irregular pore network model for convective drying and resulting damage of particle aggregates. *Chemical Engineering Science*, 75:267–278.

Kono, H. O., Chiba, S., Ells, T., and Suzuki, M. (1986). Characterization of the emulsion phase in fine particle fluidized beds. *Powder Technology*, 48(1):51–58.

Kósa, L. and Verba, A. (2001). Rheologic and flow properties of fluidised bulk solids. In *Handbook of Conveying and Handling of Particulate Solids*, volume 10 of *Handbook of powder technology*, pages 561–567. Elsevier.

Kramp, M., Thon, A., Hartge, E.-U., Heinrich, S., and Werther, J. (2012). Carbon Stripping - A Critical Process Step in Chemical Looping Combustion of Solid Fuels. *Chemical Engineering & Technology*, 35(3):497–507.

Kunii, D., Levenspiel, O., and Brenner, H. (2013). *Fluidization Engineering*. Elsevier Science, Burlington, 2nd edition.

Langmuir, I. (1908). The velocity of reactions in gases moving through heated vessels and the effect of convection and diffusion. *Journal of the American Chemical Society*, 30(11):1742–1754.

Langmuir, I. (1932). Surface chemistry. *Chem. Rev*, 13(2):147–191.

Langrish, T. and Kockel, T. K. (2001). The assessment of a characteristic drying curve for milk powder for use in computational fluid dynamics modelling. *Chemical Engineering Journal*, 84(1):69–74.

Laurindo, J. B. and Prat, M. (1998). Modeling of Drying in Capillary Porous Media: A Discrete Approach. *Drying Technology*, 16(9-10):1769–1787.

Lawrence Livermore National Laboratory (2020). SUNDIALS: SUite of Nonlinear and DIfferential/ALgebraic Equation Solvers.

Le, K. H., Tran, T. T. H., an Nguyen, N., and Kharaghani, A. (2020). Multiscale Modeling of Superheated Steam Drying of Particulate Materials. *Chemical Engineering & Technology*, 43(5):913–922.

Lee, J.-R., Lee, K.-S., Hasolli, N., Park, Y.-O., Lee, K.-Y., and Kim, Y.-H. (2020). Fluidization and mixing behaviors of Geldart groups A, B and C particles assisted by vertical vibration in fluidized bed. *Chemical Engineering and Processing: Process Intensification*, 149:107856.

Lehmann, S. E., Buchholz, M., Jongsma, A., Innings, F., and Heinrich, S. (2021). Modeling and Flowsheet Simulation of Vibrated Fluidized Bed Dryers. *Processes*, 9(1):52.

Lehmann, S. E., Hartge, E.-U., Jongsma, A., deLeeuw, I.-M., Innings, F., and Heinrich, S. (2019). Fluidization characteristics of cohesive powders in vibrated fluidized bed drying at low vibration frequencies. *Powder Technology*, 357:54–63.

Lehmann, S. E., Oesau, T., Jongsma, A., Innings, F., and Heinrich, S. (2020). Material Specific Drying Kinetics in Fluidized Bed Drying under Mechanical Vibration using the Reaction Engineering Approach. *Advanced Powder Technology*, 31(12):4699–4713.

Lettieri, P., Newton, D., and Yates, J. G. (2002). Homogeneous bed expansion of FCC catalysts, influence of temperature on the parameters of the Richardson–Zaki equation. *Powder Technology*, 123(2-3):221–231.

Levenspiel, O. (1999). *Chemical reaction engineering*. Wiley, Hoboken, NJ, 3. ed. edition.

Levenspiel, O. (2014). *Tracer technology: Modeling the flow of fluids*, volume 96 of *Fluid mechanics and its applications*. Springer, New York, NY, softcover repr. of the hardcover 1. ed. 2012 edition.

Liu, P., LaMarche, C. Q., Kellogg, K. M., and Hrenya, C. M. (2016). Fine-particle defluidization: Interaction between cohesion, Young's modulus and static bed height. *Chemical Engineering Science*, 145:266–278.

MacMullin, R. B. and Weber, M. (1935). The theory of short-circuiting in continuous flow mixing vessels in series and the kinetics of chemical reactions in such systems. *Transactions of AIChE*, 31(2):409–458.

Madhiyanon, T., Phila, A., and Soponronnarit, S. (2009). Models of fluidized bed drying for thin-layer chopped coconut. *Applied Thermal Engineering*, 29(14-15):2849–2854.

Martin, H. (2010). Heat Transfer in Fluidized Beds. In *VDI heat atlas*, VDI-Buch, pages 1300–1309. Springer-Verlag Berlin Heidelberg, Berlin.

Maurer, S., Gschwend, D., Wagner, E. C., Schildhauer, T. J., van Ommen, J. R., Biollaz, S. M., and Mudde, R. F. (2016). Correlating bubble size and velocity distribution in bubbling fluidized bed based on X-ray tomography. *Chemical Engineering Journal*, 298:17–25.

Mawatari, Y., Hamada, Y., Yamamura, M., and Kage, H. (2015). Flow Pattern Transition of Fine Cohesive Powders in a Gas-Solid Fluidized Bed under Mechanical Vibrating Conditions. *Procedia Engineering*, 102:945–951.

Mawatari, Y., Ikegami, T., Tatemoto, Y., and Noda, K. (2003). Prediction of Agglomerate Size for Fine Particles in a Vibro-fluidized Bed. *Journal of Chemical Engineering of Japan*, 36(3):277–283.

Mawatari, Y., Koide, T., Tatemoto, Y., Takeshita, T., and Noda, K. (2001). Comparison of three vibrational modes (twist, vertical and horizontal) for fluidization of fine particles. *Advanced Powder Technology*, 12(2):157–168.

McLaren, C. P., Metzger, J., Boyce, C. M., and Müller, C. R. (2021). Reduction in minimum fluidization velocity and minimum bubbling velocity in gas-solid fluidized beds due to vibration. *Powder Technology*, 382:566–572.

Meili, L., Daleffe, R. V., and Freire, J. T. (2012). Fluid Dynamics of Fluidized and Vibrofluidized Beds Operating with Geldart C Particles. *Chemical Engineering & Technology*, 35(9):1649–1656.

Mezhericher, M., Levy, A., and Borde, I. (2010). Theoretical Models of Single Droplet Drying Kinetics: A Review. *Drying Technology*, 28(2):278–293.

Molerus, O. (1982). Interpretation of Geldart's type A, B, C and D powders by taking into account interparticle cohesion forces. *Powder Technology*, 33(1):81–87.

Mortier, S. T. F. C., de Beer, T., Gernaey, K. V., Remon, J. P., Vervaet, C., and Nopens, I. (2011). Mechanistic modelling of fluidized bed drying processes of wet porous granules: a review. *European journal of pharmaceutics and biopharmaceutics: official journal of Arbeitsgemeinschaft fur Pharmazeutische Verfahrenstechnik e.V*, 79(2):205–225.

Mujumdar, A. S. and Devahastin, S. (2003). Application for fluidized bed drying. In Yang, W.-c., editor, *Handbook of fluidization and fluid-particle systems*, volume chapter 18 of *Chemical industries*, pages 469–484. Marcel Dekker, New York.

Murti, R. A., Paterson, A., Pearce, D. L., and Bronlund, J. E. (2010). The influence of particle velocity on the stickiness of milk powder. *International Dairy Journal*, 20(2):121–127.

Nam, C. H., Pfeffer, R., Dave, R. N., and Sundaresan, S. (2004). Aerated vibrofluidization of silica nanoparticles. *AIChE Journal*, 50(8):1776–1785.

Nauman, E. B. (2008). Residence Time Theory. *Industrial & Engineering Chemistry Research*, 47(10):3752–3766.

Nilsson, L. and Wimmerstedt, R. (1988). Residence time distribution and particle dispersion in a longitudinal-flow fluidized bed. *Chemical Engineering Science*, 43(5):1153–1160.

Noda, K., Mawatari, Y., and Uchida, S. (1998). Flow patterns of fine particles in a vibrated fluidized bed under atmospheric or reduced pressure. *Powder Technology*, 99(1):11–14.

Palzer, S. (2005). The effect of glass transition on the desired and undesired agglomeration of amorphous food powders. *Chemical Engineering Science*, 60(14):3959–3968.

Palzer, S. (2010). Predicting drying kinetic and undesired agglomeration during drying of granules containing amorphous water-soluble substances in a continuous horizontal fluid bed dryer. *Powder Technology*, 201(3):201–212.

Park, W. H., Kang, W. K., Capes, C. E., and Osberg, G. L. (1969). The properties of bubbles in fluidized beds of conducting particles as measured by an electroresistivity probe. *Chemical Engineering Science*, 24(5):851–865.

Penn, A., Boyce, C. M., Pruessmann, K. P., and Müller, C. R. (2020). Regimes of jetting and bubbling in a fluidized bed studied using real-time magnetic resonance imaging. *Chemical Engineering Journal*, 383:123185.

Perazzini, H., Freire, F. B., and Freire, J. T. (2017). The influence of vibrational acceleration on drying kinetics in vibro-fluidized bed. *Chemical Engineering and Processing: Process Intensification*, 118:124–130.

Pietsch, S. (2018). *Fluidization behavior and liquid injection in three-dimensional prismatic spouted beds*. Dissertation, Technische Universität Hamburg.

Pietsch, S., Kieckhefen, P., Müller, M., Schönherr, M., Kleine Jäger, F., and Heinrich, S. (2018). Novel production method of tracer particles for residence time measurements in gas-solid processes. *Powder Technology*, 338:1–6.

Pietsch, S., Schönherr, M., Kleine Jäger, F., and Heinrich, S. (2020). Measurement of Residence Time Distributions in a Continuously Operated Spouted Bed. *Chemical Engineering & Technology*, 43(5):804–812.

Pisecký, J. (2012). *Handbook of milk powder manufacture*. GEA Niro, Soeborg, 2nd ed. edition.

Puettmann, A., Hartge, E.-U., and Werther, J. (2012). Application of the flowsheet simulation concept to fluidized bed reactor modeling. Part I. Development of a fluidized bed reactor simulation module. *Chemical Engineering and Processing: Process Intensification*, 60:86–95.

Puncochar, M., Ruzicka, M. C., and Simcik, M. (2016). Bubble swarm rise velocity in fluidized beds. *Chemical Engineering Science*, 152:84–94.

Putranto, A., Chen, X. D., Devahastin, S., Xiao, Z., and Webley, P. A. (2011a). Application of the reaction engineering approach (REA) for modeling intermittent drying under time-varying humidity and temperature. *Chemical Engineering Science*, 66(10):2149–2156.

Putranto, A., Chen, X. D., and Webley, P. A. (2011b). Modeling of Drying of Food Materials with Thickness of Several Centimeters by the Reaction Engineering Approach (REA). *Drying Technology*, 29(8):961–973.

Putranto, A., Xiao, Z., Chen, X. D., and Webley, P. A. (2011c). Intermittent Drying of Mango Tissues: Implementation of the Reaction Engineering Approach. *Industrial & Engineering Chemistry Research*, 50(2):1089–1098.

Raganati, F., Chirone, R., and Ammendola, P. (2018). Gas–solid fluidization of cohesive powders. *Chemical Engineering Research and Design*, 133:347–387.

Reay, D. (1978). Particle residence time distributions in "plug flow" fluid bed dryers. In Mujumdar, A. S., editor, *Proceedings of 1st International Drying Symposium, Montreal, Canada 1978*, pages 136–144, Princton, NJ. Sci Press.

Retsch Technology GmbH (2017). Application Note CamSizer XT - Food. *Manufacturere Application Note*, 1(1):1–4.

Richardson, J. F. and Zaki, W. N. (1954). The sedimentation of a suspension of uniform spheres under conditions of viscous flow. *Chemical Engineering Science*, 3(2):65–73.

Rüdisüli, M., Schildhauer, T. J., Biollaz, S. M., and van Ommen, J. R. (2012). Bubble characterization in a fluidized bed by means of optical probes. *International Journal of Multiphase Flow*, 41:56–67.

Satija, S. and Zucker, I. L. (1986). Hydrodynamics of Vibro-Fluidized Beds. *Drying Technology*, 4(1):19–43.

Skorych, V., Das, N., Dosta, M., Kumar, J., and Heinrich, S. (2019). Application of Transformation Matrices to the Solution of Population Balance Equations. *Processes*, 7(8):535.

Skorych, V., Dosta, M., Hartge, E.-U., and Heinrich, S. (2017). Novel system for dynamic flowsheet simulation of solids processes. *Powder Technology*, 314:665–679.

Skorych, V., Dosta, M., and Heinrich, S. (2020). Dyssol—An open-source flowsheet simulation framework for particulate materials. *SoftwareX*, 12:100572.

Stakić, M. and Urošević, T. (2011). Experimental study and simulation of vibrated fluidized bed drying. *Chemical Engineering and Processing: Process Intensification*, 50(4):428–437.

Stieß, M. (1995). *Mechanische Verfahrenstechnik 1*. Springer-Lehrbuch. Springer, Berlin and Heidelberg, 2. edition.

Surasani, V. K., Metzger, T., and Tsotsas, E. (2008). Consideration of heat transfer in pore network modelling of convective drying. *International Journal of Heat and Mass Transfer*, 51(9-10):2506–2518.

Surface Measurement Systems Ltd. (2015). DVS Resolution - Product Brochure: Product Brochure.

Tsotsas, E. (2010). Heat Transfer from a Wall to Stagnantand Mechanically Agitated Beds. In *VDI heat atlas*, VDI-Buch, pages 1311–1326. Springer-Verlag Berlin Heidelberg, Berlin.

Tsotsas, E., Metzger, T., Gnielinski, V., and Schlünder, E.-U. (2000). *Ullmann's Encyclopedia of Industrial Chemistry: Drying of Solid Materials*. Wiley-VCH Verlag GmbH & Co. KGaA, Weinheim, Germany.

Tsotsas, E. and Mujumdar, A. S. (2011). *Modern Drying Technology, Experimental Techniques*. Modern Drying Technology. John Wiley & Sons, Hoboken.

Valverde, J. M. and Castellanos, A. (2006). Effect of vibration on agglomerate particulate fluidization. *AIChE Journal*, 52(5):1705–1714.

Valverde, J. M. and Castellanos, A. (2008). A modified Richardson–Zaki equation for fluidization of Geldart B magnetic particles. *Powder Technology*, 181(3):347–350.

van Meel, D. A. (1958). Adiabatic convection batch drying with recirculation of air. *Chemical Engineering Science*, 9(1):36–44.

van Ommen, J. R. (2009). Reshaping the Structure of Fluidized Beds. *Chemical Engineering Progress*, 105(7):49–57.

van Ommen, J. R., van der Schaaf, J., Schouten, J. C., van Wachem, B. G., Coppens, M.-O., and van den Bleek, C. M. (2004). Optimal placement of probes for dynamic pressure measurements in large-scale fluidized beds. *Powder Technology*, 139(3):264–276.

Vollmari, K. and Kruggel-Emden, H. (2018). Numerical and experimental analysis of particle residence times in a continuously operated dual-chamber fluidized bed. *Powder Technology*, 338:625–637.

Vuataz, G. (2002). The phase diagram of milk: a new tool for optimising the drying process. *Le Lait*, 82(4):485–500.

Wadell, H. (1933). Sphericity and Roundness of Rock Particles. *The Journal of Geology*, 41(3):310–331.

Wang, T.-J., Jin, Y., Tsutsumi, A., Wang, Z., and Cui, Z. (2000). Energy transfer mechanism in a vibrating fluidized bed. *Chemical Engineering Journal*, 78(2-3):115–123.

Wang, Y., Wang, T.-J., Yang, Y., and Jin, Y. (2002). Resonance characteristics of a vibrated fluidized bed with a high bed hold-up. *Powder Technology*, 127(3):196–202.

Wei, L., Lu, Y., Jiang, G., Hu, J., and Zhu, J. (2017). Unsteady-state bubble dynamic wave velocity of gas–solid bubbling fluidized bed. *Chemical Engineering Research and Design*, 126:255–264.

Wensrich, C. M. and Collard, A. (2006). Resonant and non-linear behaviour in vibrationally fluidised beds. *Powder Technology*, 166(1):30–37.

Werther, J. (1974). Influence of the Bed Diameter on the Hydrodynamics of Gas Fluidized Beds. *AIChE Symposium Series*, 70(141):53–62.

Werther, J., Hartge, E.-U., and Heinrich, S. (2014). Fluidized-Bed Reactors - Status and Some Development Perspectives. *Chemie Ingenieur Technik*, 86(12):2022–2038.

Werther, J. and Molerus, O. (1973). The local structure of gas fluidized beds —I. A statistically based measuring system. *International Journal of Multiphase Flow*, 1(1):103–122.

Wiesendorf, V. (2000). *The capacitance probe - a tool for flow investigations in gas-solids fluidization systems*. Dissertation, Hamburg University of Technology, Hamburg.

Wiesendorf, V. and Werther, J. (2000). Capacitance probes for solids volume concentration and velocity measurements in industrial fluidized bed reactors. *Powder Technology*, 110(1-2):143–157.

Winkler, F. (1922). Verfahren zur Herstellung von Wassergas. *Patent*, DE437970.

Wirth, K.-E. (2010). Flow Patterns and Pressure Drop in Fluidized Beds. In *VDI heat atlas*, VDI-Buch, pages 1197–1206. Springer-Verlag Berlin Heidelberg, Berlin.

Wytrwat, T., Yazdanpanah, M., and Heinrich, S. (2020). Bubble Properties in Bubbling and Turbulent Fluidized Beds for Particles of Geldart's Group B. *Processes*, 8(9):1098.

Xu, C. and Zhu, J. (2005). Experimental and theoretical study on the agglomeration arising from fluidization of cohesive particles—effects of mechanical vibration. *Chemical Engineering Science*, 60(23):6529–6541.

Xu, C. and Zhu, J. (2006). Parametric study of fine particle fluidization under mechanical vibration. *Powder Technology*, 161(2):135–144.

Zhang, C., Li, P., Lei, C., Qian, W., and Wei, F. (2018). Experimental study of non-uniform bubble growth in deep fluidized beds. *Chemical Engineering Science*, 176:515–523.

Zhang, Y., Zhang, X., Zhao, Y., Lv, G., Yang, X., Wang, Z., Duan, C., and Dong, L. (2020). Bubble growth obtained from pressure fluctuation in vibration separation fluidized bed using wavelet analysis. *Advanced Powder Technology*, 31(8):3287–3296.

Zhao, P., Zhao, Y., Luo, Z., Chen, Z., Duan, C., and Song, S. (2014). Effect of operating conditions on drying of Chinese lignite in a vibration fluidized bed. *Fuel Processing Technology*, 128:257–264.

Zhou, T., Ogura, H., Yamamura, M., and Kage, H. (2004). Bubble Motion Pattern and Rise Velocity in Two-dimensional Horizontal and Vertical Vibro-fluidized Beds. *The Canadian Journal of Chemical Engineering*, 82(2):236–242.

Zou, Z., Li, H., and Zhu, Q.-s. (2011). The bubbling behavior of cohesive particles in the 2D fluidized beds. *Powder Technology*, 212(1):258–266.

# Curriculum Vitae

| | |
|---|---|
| Name | Sören Ernst Lehmann |
| Born | 21.01.1990 in Buchholz i.d.N. |

**Education**

| | |
|---|---|
| 1996 - 2000 | Grundschule Hittfeld |
| 2000 - 2006 | Orientierungsstufe Hittfeld |
| 2006 - 2009 | Gymnasium Hittfeld, Abitur |

**Academic History**

| | |
|---|---|
| 10/2009 - 10/2012 | Study of Energy and Environmental Engineering (B.Sc.), Hamburg University of Technology, grade 2.0 |
| 10/2012 - 02/2016 | Study of Energy and Environmental Engineering (M.Sc.), Hamburg University of Technology, grade 1.3 |
| 03/2013 - 11/2013 | Study of Geothermal Energy Technology (PGCert), University of Auckland, New Zealand, grade 1.7 |
| 10/2014 - 12/2014 | Research project, University of Canterbury, Christchurch, New Zealand, grade 1.3 |
| 06/2016 - 12/2020 | Scientific researcher and PhD student at the Institute of Solids Process Engineering and Particle Technology, Hamburg University of Technology.<br>Supervisor: Prof. Dr.-Ing. habil. Dr. h.c. Stefan Heinrich |

**Career**

| | |
|---|---|
| Since 04/2021 | Project Manager at Glatt Pharmaceutical Services GmbH & Co.KG |

www.ingramcontent.com/pod-product-compliance
Ingram Content Group UK Ltd.
Pitfield, Milton Keynes, MK11 3LW, UK
UKHW061658200726
13853UKWH00013B/2466